EXPLORING THE
RIVER FOWEY

JOHN NEALE

AMBERLEY

For my dear sisters- and brothers-in-law, Dorothy and Roy Neale and Barbara and Arthur Masters, and my four nephews – Steven, Ian, Andrew and Stuart – who all live on Bodmin Moor, within the proverbial stone's throw of the infant River Fowey and probably know its upper reaches better than I do!

First published 2013

Amberley Publishing
The Hill, Stroud
Gloucestershire, GL5 4EP

www.amberley-books.com

Copyright © John Neale 2013

The right of John Neale to be identified as the Author of this work has been asserted in accordance with the Copyrights, Designs and Patents Act 1988.

All rights reserved. No part of this book may be reprinted or reproduced or utilised in any form or by any electronic, mechanical or other means, now known or hereafter invented, including photocopying and recording, or in any information storage or retrieval system, without the permission in writing from the Publishers.

British Library Cataloguing in Publication Data.
A catalogue record for this book is available from the British Library.

ISBN 978 1 4456 0648 4

Typeset in 10pt on 12pt Sabon.
Typesetting and Origination by Amberley Publishing.
Printed in the UK.

Contents

	Acknowledgements	4
	Introduction	5
	Map of River	7
1	The Infant Stream	9
2	The Peaty River	24
3	King Doniert Country	38
4	Treverbyn to the Glynn Valley and Lanhydrock	55
5	Braddock, Boconnoc House to Lostwithiel	69
6	Lostwithiel and Further Downstream	80
7	Downstream to Fowey	104
8	Bodinnick, Lanteglos-by-Fowey and Polruan	117
	Further Reading	128

Acknowledgements

My thanks go to: Miss Angela Broom and her volunteer helpers at the Courtney Library, Royal Institution Cornwall, Truro; the archivists at the Cornwall County Record Office, Truro; Miss Kim Cooper and the staff at the Cornwall Centre, Redruth; Neil Williams at the Photographic Library, Cornwall Centre, Redruth; Mark Thackeray at the Cornwall County Reference Library, Threemilestone, Truro; the staff at the Cornwall County libraries in Launceston, Bodmin, Liskeard and Fowey; the Cornwall Archaeological Unit (Historic Environmental Services), Truro; the staff at the Fowey Harbour Master's office; the staff at Clemens Photography, Bodmin; the curator, Miss Tremar Menendez, and stewards at Lostwithiel Museum; the curator, Miss Helen Luther, and stewards at Fowey Museum; Revd Louise Courteny, Lanteglos-by-Fowey; Isabel Pickering, Fowey; Jane Talbot Smith at Higher Harrowbridge Saddlery; Revd Andrew Balfour, St Neot; Rose Wilton, St Neot; Mr Andrew Blunt, St Neot; Tony Lucas, picture postcard collector and dealer, Saltash; the picture postcard publishers Overland, Ellis, Valentine, Frith, Argyll, etc.; *Picture Postcard Magazine* and its subscribers; Elizabeth Fortescue, Boconnoc; Robert Tremain; Peter Davies, Bodmin; Robert Evans; Colin Barratt; Roger Hancock of the Trigg Morris Men, Bodmin; postmen Phil Sullivan and Ivan Flower, without whose directions I would never have found some bridges over the River Fowey; Mr Jasper Bagshawe; Sheila Lee; Miss Chloe Johnson; Raymond Hunkin; Tony and Sonia Hillman, Launceston; the files of the *East Cornwall Times*, Liskeard; the files of the *Cornish Guardian*, Bodmin; the files of the *West Briton*; the files of the *Royal Cornwall Gazette*; the files of the *Cornish and Devon Post* and *Launceston Weekly News*; the files of the *Western Morning News*; those photographers who failed to mark their work.

My unreserved apologies to anyone who has been inadvertently overlooked.

Introduction

Having written two previous books about rivers, *Discovering the River Tamar* and *Following the Camel*, to be asked to write a third book, featuring a river of my choice, came as something of an agreeable surprise.

Compared with the Severn, Tyne, Humber, or Mersey and rivers in other parts of the country, West Country rivers and Cornish rivers in particular are reckoned, in general, to be on the smaller scale. This is something of a myth because no matter what their size or how they are viewed, in reality, they lack nothing in comparison or in interest or beauty and their natural deep harbours and estuaries are unrivalled. But which river to write about? Initially this presented itself as something of a dilemma. Ought it be a Devonshire or a Cornish river? On first thought, should it be the Dart, Taw, Exe or Torridge? For one reason or another each was dismissed. The remaining options were the Helford, the Fal or the Fowey. Geography too came into the equation and as one of my favourite areas of north Cornwall is Bodmin Moor, which vastness gives birth to the River Fowey, the choice, to all intents and purposes, was ready-made. It has to be the Fowey, which among the rivers of Cornwall, ranks in the first division and has its own magnetism.

Someone once said that rivers are like people: each has its own idiosyncrasies and ambitions. But what is the ambition of the River Fowey? From the outset its goal is to become tidal and to greet the sea. For most of the time everyone generally loves the River Fowey; at times, according to the season, it is quiet and affable and minding its own business, but at other times its mood changes and it becomes rude, boisterous and belligerent, overruling its general confines, sweeping away everything in its path and causing untold damage.

Nature seems to have carved the River Fowey up into three roughly equal sections. At first the fledgeling stream makes its way tentatively through rough, sparsely populated moorland; here, for some distance its banks are lined with scrub and wind-bent trees. Later pasture-land reaches its banks and later still it is hemmed in by dense woodland; then, in its lower reaches, it blossoms and shows its glory as it winds and widens to display its harbour, its estuary and the open sea.

The River Fowey rises, so Leland wrote in 1538, 'in a very wagmire in the side of a hill'. The hill is Brown Willy, Cornwall's highest point. The actual source of the Fowey is Fowey Venton or Fowey Well, which spot was once marked by the chapel of St Peter Fawe, of which today there is little or no trace.

In the eighteenth century, what is now known as Bodmin Moor was known as Fowey Moor. Crossing the moor was a risky business; ordinary folk avoided the experience and those who did venture did so on foot or used pack-horses, following the rough tracks, and risked being robbed every mile of the way. The Courts of Assize were held in Launceston as the circuit judges are reputed to have shunned travelling over the moor for the same reason. It was the building of the new turnpike roads across the moor during the nineteenth century which opened up this part of Cornwall. The new roads gave easier access for coaches and horses, and the Assize courts were moved to Bodmin. Around this time Bodmin was growing rapidly and soon became the largest centre of population near to the geographical centre of Cornwall. It is this and the influence of early travellers that is believed to have brought about the moor's change of name. Later, the romantic name Fowey Moor was dropped in favour of what was believed to be more appropriate: Bodmin Moor. The fact that originally the River Fowey gave its name to the moor and not Bodmin town seems to have been perhaps conveniently brushed aside by those who had the say at the time!

The eminent writer Charles Henderson wrote that the Devonians have not behaved so ungenerously to their Dart as to change the name of Dartmoor, but the Cornish changed the name of Fowey Moor to the meaningless Bodmin Moor! There is a growing number today who agree with this sentiment and believe that the old name ought to have been retained; after all, they argue, the Exe gives its name to Exmoor and the Dart to Dartmoor. Perhaps after all these decades it is an impossible pipe-dream to bring any reversal of name to fruition.

The River Fowey, for part of its roughly 30-mile southerly course, forms the southern boundary of Bodmin Moor. Later the growing river veers west as if fighting to discover a way to reach the sea before redirecting itself southwards again and achieving its ambition. Along its way the Fowey plays host to a handful of tributaries, the stream flowing by Trenant, the Loveny (pronounced Love-enny), the Warleggan River and the Cardinham Water, each careering down from Bodmin Moor to join the bigger river. For some strange reason, two of these names have changed over the years. The Loveny is now generally known as the St Neot River while the Bedalder has become the Warleggan River, which rises near Hawk's Tor in Blisland parish. It is curious why this should have happened, two beautiful, romantic-sounding river names dropped for more mundane substitutes. Surely it must be largely due to the mapmakers over the years and not to the local people! Obviously this is a topic which will give rise to heated discussion in the bars of several local hostelries!

Someone once described the River Thames as being liquid history; perhaps the same may be said of the River Fowey, albeit on a different level. There are many things of interest to look out for on the banks of the River Fowey, which are flanked with history. Explorers will discover ancient towns, old houses and a string of fascinating villages with ancient churches steeped in history, and most have a tale to tell or something to reveal and all are focal points in the hinterlands of the River Fowey. Along the way a rich mosaic of old inns, chapels, bridges, industrial archaeology, slate caverns, a holy well, a castle, battle sites, ancient inscribed stones and earthworks are all encountered. Stories of ghosts, an oddball vicar and legends are retold as well as modern mysteries, all adding to the richness of the territory of the River Fowey.

It is fair to suggest that probably no other Cornish river can boast stronger literary associations than the River Fowey and its watershed. Some of these authors enjoy enviable worldwide reputations, others are well-known and some less so, but among the glitterati are Daphne Du Maurier, George Borrow, Leo Walmsley, Mabel Lucie Attwell, Kenneth Grahame, Denys Val Baker, and 'Q', Sir Arthur Quiller Couch, arguably Cornwall's and certainly Fowey's own great man of letters.

The Fowey is the thread which connects all these facets of the river scene – people, places and events – in an unbroken time-line, until it eventually spreads itself to flow majestically between Bodinnick and Polruan on the one side and Fowey on the other to be embraced by the open sea.

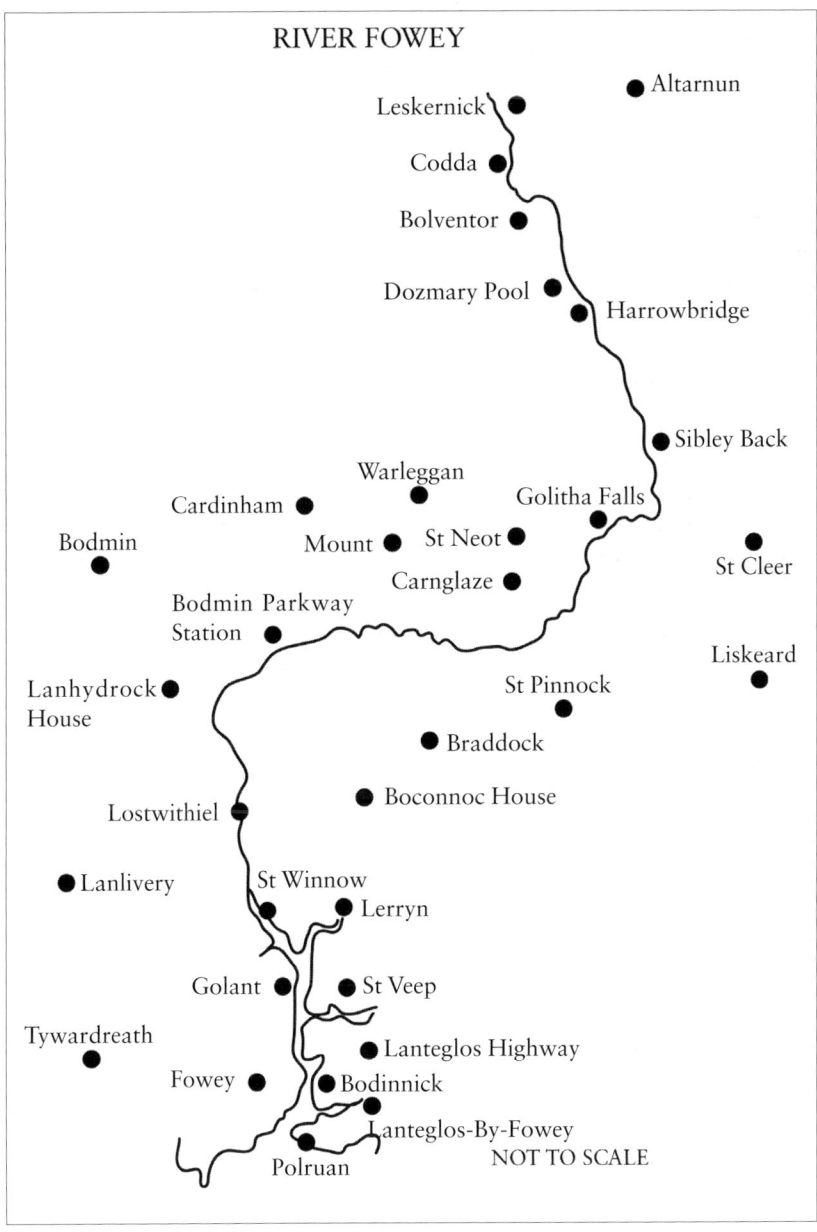

The Jamaica Inn sign is photographed thousands of times every year by people from all over the world and it is believed to be the most photographed inn sign in the country. This picture was taken in January 2013, and the photographer was not alone! Seemingly some of Daphne Du Maurier's Polish fans have discovered the road from Warsaw to Jamaica Inn!

I

The Infant Stream

In the remotest part of Bodmin Moor, the pockmarked landscape of tumuli – Bronze Age hut circles and the spoil heaps left by tin streamers, mounds which have now grassed over – and the attendant water leats gives strong evidence of human occupation since earliest times. From here, flanked by Roughtor, Brown Willy and Kilmar, the streamlets that will soon grow and form the River Fowey emerge almost furtively from the marshy cradle.

This part of Bodmin Moor or Fowey Moor or, originally, Fawimor is mainly a vast open space, and one of the most isolated parts of the moor. This is where the River Fowey becomes recognisable as a river, flowing southwards – the only Cornish river to do so for much of its course – to start its fascinating journey as it carves its way through the scrub- and gorse-jammed valley, between two of the most isolated farms, Leskernick on the one side and Codda on the other. Soon, having only travelled a short distance, the Fowey demands a footbridge.

The first dwelling on the infant Fowey is Leskernick farm in its remote setting. At one time the farm was occupied by the Rich family. Later the house stood empty for a number of years, but it is currently occupied again.

Altarnun is one of Bodmin Moor's more picturesque villages, where sturdy granite buildings line the main street and a small, ancient single-file stone bridge spans the Penpont Water and the parish church is aptly called the Cathedral of the Cornish Moors. In early spring the scene is 'set off' by drifts of snowdrops over the churchyard hedge and in the field opposite, looking as if they had been thickly painted on by someone using a giant paintbrush.

Over the years Altarnun entered the Britain in Bloom contest, but the work involved became too much for a small organising committee and a handful of helpers and it was decided to enter the Best Kept Village contest, run by the Council for the Protection of Rural England. Altarnun entered for twelve years, winning five times, which at one period included a 'hat-trick' of wins. In 1987 villagers were angered because they were asked to stand down when they were on a winning streak in order to give other villages a chance of winning. Needless to say this caused much debate in the community and irate letters were sent to the local newspapers. However, success returned when in 1999 Altarnun fought off stiff competition from eighteen other villages by chalking up top points in each section – absence of litter, maintenance, tidiness and use of space – to win a fourth plaque.

The parish church is the focal point of one of the largest parishes, whose sprawling acres vie for the favour of the River Fowey with those of Bolventor. Altarnun church stands on high ground at one end of the village and is dedicated to St Nonna, the mother of St David.

Brown Willy, Cornwall's highest hill, together with Roughtor and Kilmar – fondly called 'The Three Monarchs' by George Ellis, the doyen of photographers from Bodmin – seems to be guarding the remote part of Bodmin Moor where the River Fowey rises.

Leskernick farm, Bolventor, the first dwelling on the infant River Fowey in its remote Bodmin Moor setting. At one time Leskernick was occupied by the Rich family. Later the house stood empty for a number of years, but it is currently occupied again. (Photograph courtesy Peter Davies collection)

Codda Farm, nestling in its moorland hollow, is one of the most interesting among a clutch of historic houses on Bodmin Moor.

Altarnon, twinned with Gueltas in Brittany, is one of a handful of picturesque villages on Bodmin Moor. On several occasions Altarnon has been voted the UK's best-kept village, as shown on dated wall plaques at Penpont.

The church tower soars to 100 feet and was mainly built of moorstone during the fifteenth century. In June 1986 the church tower was struck by lightning during a horrendous thunderstorm, and a pinnacle weighing about 3 tons was dislodged and later came crashing down through the roof, destroying a sixteenth-century bench-end and damaging the floor. Of the nearly eighty surviving sixteenth-century bench-ends, all are the workmanship of Robart Daye, believed to be a local man. They show typical country scenes such as men playing musical instruments – one in particular shows a viellie, an early form of violin – as well as jesters and dancers. The font too is interesting as it was kept outdoors during the building of the present church. At one period it was painted black and red, slight traces of which can still be seen. The rood screen, which stretches across the width of the church, is sadly not complete but the Jacobean altar rails were believed to be the work of John Gard, a carpenter from Launceston.

The disaster rallied the villagers, who decided that church work must carry on. Tons of rubble and smashed woodwork were cleared and a thick coat of dust throughout the church was removed. Within a few days, a memorial service was held for two local young men who had been killed in a recent farm tragedy, a wedding temporarily removed to nearby North Hill church was brought back to Altarnun, and a flower festival went ahead as planned. It was estimated at the time that the repairs would cost upwards of £20,000. Apparently Altarnun church was struck by lightning during the eighteenth century when a vestry window and a wall were destroyed.

Altarnon church, affectionately called the Cathedral of the Cornish Moors, stands sentinel over the lower end of the village. Digory and Elizabeth Isbell, who gave John Wesley food and shelter in their cottage at Trewint, are buried here. There are also tombstones carved by Neville Northey Burnard, arguably one of Altarnon's famous sons.

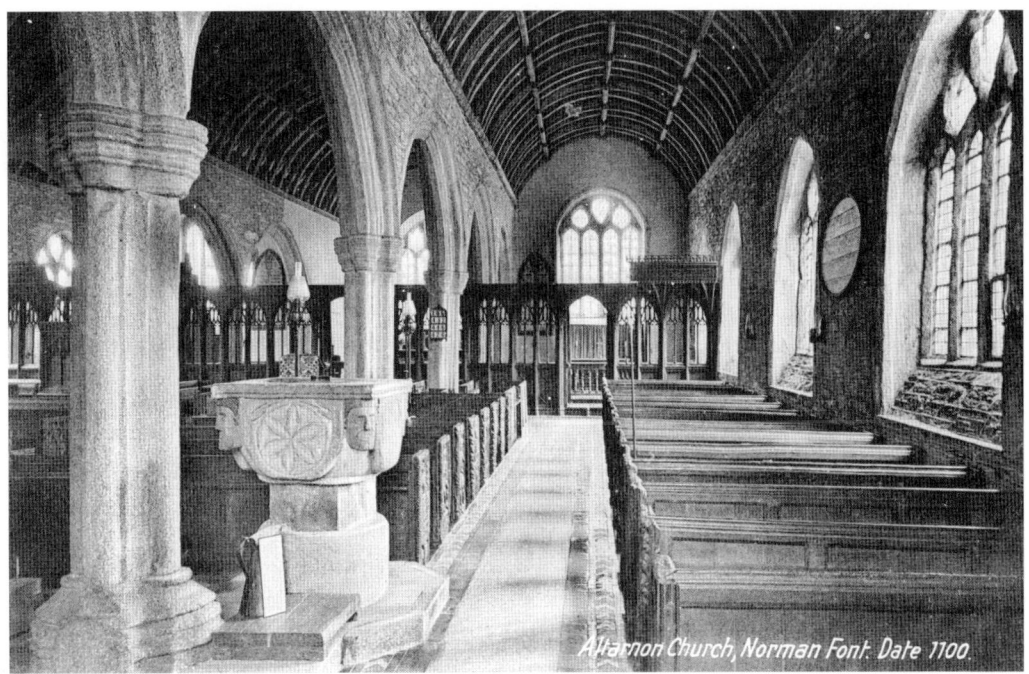

The interior of Altarnon church boasts a fascinating collection of bench-ends showing various subjects and country scenes. The font at the back of the church and the wooden screen, which stretches the full width of the building, are both noteworthy.

Near the church porch are some interesting tombstones carved by Altarnun's famous sculpture son, Neville Northey Burnard. The son of a stonemason, he was born in 1818 just a few hundred yards away at Penpont Mill. Burnard's mother kept a Dame School and she made straw hats in her spare time. Seemingly at an early age, Neville showed an aptitude for drawing and carving birds, flowers and animals on slate. A tombstone to Burnard's father George near the church porch shows a pheasant in full flight carved by Burnard with nothing but a sharpened nail when he was fourteen years of age. Fortunately his name, though often hidden by grass, is still discernible at the bottom of the tombstone.

Penpont Mill, where Burnard was born, boasts a commemorative plaque and the early Methodist Meeting House, which carries a bust of John Wesley carved by Burnard, is extant. Sadly, after the early loss of his daughter Burnard turned to drink and he eventually died penniless in Redruth workhouse.

Five Lanes, a short distance from Altarnun, is a growing community; since the closure of the cattle market the site has been used for housing. The village is centred around the Kings Head Hotel.

At Trewint, a good gunshot away, is Wesley Cottage, the historic centre of Methodism in the area. It was in 1744 that two of John Wesley's agents, John Downes and John Nelson, wet, tired and hungry, called unexpectedly at the door of a small cottage, the home of Digory and Elizabeth Isbell. Unfazed, Elizabeth invited the two strangers into her home and gave them sustenance. Shortly afterwards Digory, who was a devout believer, was reading his Bible

This image shows the old single-file bridge over the Penpont water at Altarnon where there was once a ford on the lower side of the old bridge, which has now been spanned by a modern bridge.

Penpont Mill, Altarnon, the birthplace in 1818 of Neville Northery Burnard, who found fame as a sculptor in London. He later returned to Cornwall and in 1878 died a pauper in the workhouse at Redruth.

when he came across the story of the Shunnamite woman who built a prophets room in which to entertain a man of God. Digory was obviously much affected by the story and determined to do the same. He added two small rooms, one up and one down, to his cottage.

At one time Trewint was a Methodist stronghold. But with local competition from smaller chapels springing up, each with their own following, Trewint found itself somewhat sidelined, and the cottage fell into disrepair and became a roofless ruin. In the 1940s, Stanley Sowton of nearby North Hill, aware of the history of the cottage, became concerned about the building and devised a plan for rescue, and in 1950 it was opened to the public. Inside there is an exhibition of Wesleyana. For the last sixty years and more, come rain or shine, Wesley Day celebrations have been held. Significant landmark dates have been suitably commemorated. On the fortieth in 1990, a pageant, 'A Trewint Tapestry', with the cast dressed in appropriate costume, was mounted by the Reverend Tom Shaw. In later years, special attendees were Mr John Moyse, who was involved in the original restoration, and Mr Ted Isbell and his family. Mr Isbell is a direct descendant of Digory and Elizabeth Isbell. For many years the Pooley family have been involved with Wesley Cottage; Mr Lionel Pooley was secretary to the managing committee for over twenty years and Mrs Joyce Pooley was caretaker-cum-curator for many years until ill health forced her to retire. Around this time new government legislation decreed that the Methodist district could not have a voluntary caretaker, which means that the cottage can only be opened for limited periods; this has greatly affected income, and together with much-needed roof repairs means that the future of the cottage is in jeopardy. Hopefully the Methodist district will take on full responsibility.

Another Altarnun surprise is Codda, a remarkable granite-built gem and a little-known moorland dwelling. Codda is discovered less than a mile from the main A30 trunk road which slices the vast waste of Bodmin Moor. Codda is reached by a long, rough track and crossing a careering stream as it rushes through watery meadows to join the infant Fowey. Codda is a Grade II listed building. Today it is seemingly plonked down in the great wasteland on the south-facing slopes on the northern side of Bodmin Moor, tucked away in its sheltered valley underneath the roughest weather that can roar across the moor in winter. Originally Codda is thought to have been a longhouse where there was accommodation for humans and animals all under the same roof, partitioned from one another across a dividing passage. Codda is an important feature in this part of the moorland landscape. It was a sturdy farmstead in the seventeenth century and evidence to support this can be spotted in carved stones from near this date around some of the windows and doorways in the building and in some of the surviving outbuildings.

Codda is original and defies the law of averages, because what is amazing is the fact that the homestead, by some quirk of fate, possibly due in some part to its isolation, has seemingly leap-frogged the centuries, largely untouched from the eleventh to our own time. Built entirely of granite from the surrounding moorland, Codda dates from 1239 according to documentary evidence and research by some eminent historians. Other historians place the date some two centuries earlier, to a time possibly prior to the Norman Conquest.

At one time the homestead was known as Stuncodda and an Assize Roll of 1280 shows that Wulter Polit and William Kena leased it to Stephen and Joan Rogger. In May 1385, when William Fitzwauter died it was found that according to his will he held land at Stymkodda and Brownwalyng (Brown Willy) in the manor of Fawton, which land stretched from St Neot across large tracts of Bodmin Moor, including most of the parish of Altarnun. Over 100 years later in 1540, on a rent roll of Roscarrock Manor, mention is made of Stynnecodde. In 1571 Codda still had two households; for some reason an exodus away from the moor seems to have begun. Twenty years later in 1591, part of Fawton Manor was sold to Reignolde Mohun of Boconnoc and included Stevendon Codda. In the seventeenth century, when Altarnun church required money for essential maintenance and general running costs, it would seem that much like today, a parish collection was made. It is recorded that two residents of Codda, Nicholas Speare and Jane Dennis, gave *6d* (two and a half pence) each. Of the other forty-six people who contributed, only five people equalled or gave more than these two Codda residents.

The exodus from the moor seems to have continued, as on an old map of 1748 Codda is indicated as Coaddha and being one farmhouse. Another change of ownership came in the nineteenth century, when Codda was bought from Lord St Germans and a Mrs West by Francis Hearle Rodd of Trebartha, North Hill. Around this time the house was extended on the east-facing side. Later Rodd was seemingly land-grabbing wherever he could on the moor to create his new parish, known today as Bolventor. Later Codda was owned by another Francis Rodd and occupied by Richard Vosper. By the middle of the nineteenth century, Richard Bray lived at Codda and farmed numerous acres. In the twentieth century Codda lost its importance and by the 1970s the house was no longer a farmstead and the land was worked from another remote farmstead. In more recent times, a family named Jasper occupied Codda until about the early 1980s.

After this time Codda was abandoned, and in 1989 part of the roof was missing and a section of the end wall had collapsed and it was fast becoming derelict. Rescue, however, was at hand when Mr and Mrs Jasper Bagshawe purchased Codda together with a range of outbuildings and decided to rescue the farmstead. In January 1999, Mr and Mrs Bagshawe, together with English Heritage and the North Cornwall District Council (now Cornwall Council), commissioned the Cornwall Archaeological Unit (now Cornwall Historic Environmental Services) to record and interpret the property. Mr and Mrs Bagshawe did a colossal amount of necessary restoration at Codda as part of their rescue scheme. Today Codda, as a whole, is reckoned to be of both national and county importance.

Codda has a range of granite outbuildings and a quintet of goose holes. The small stone caves were built into a low hedge, each with a heavy stone sliding 'door', put in place after the geese, probably three or four, had been installed in each hole, to keep them safe, warm and dry and to protect them from predatory foxes.

A short distance away, the growing River Fowey makes its way toward Palmers Bridge, the first significant crossing over the Fowey. In early October 1880 the area was struck by a tremendous downpour of rain; the worst, so the *Launceston Weekly News* reports, for sixteen years. Great damage was done to roads at Palmers Bridge, near Bolventor, and the bridge was so severely damaged that traffic had to be stopped crossing it for several days while it was being repaired. Two months later, a report tells that a new road is being cut a little above the old bridge which will serve before the bridge is rebuilt. Prior to the new dual carriageway being built, the present Palmers Bridge was built. Legend has it that the new bridge was built to allow for the road-widening scheme to proceed. On its completion the river was diverted; it takes a distinct right-angle turn to flow under the new bridge. The old bridge was later torn down!

The Infant Stream

The River Fowey begins to gather speed as it approaches Palmers Bridge on the main A30 trunk road between Launceston and Bodmin. Immediately in front of the bridge the river abruptly changes direction and dives under the road bridge and scurries on through woodland.

Here the River Fowey begins to show its strength as it flows under the main A30 trunk road.

The toll-house was an octagonal two-storey building, with side windows looking either way. There was no hope of not being spotted and so avoiding paying the tolls. The building was demolished in March 1986 before the new dual carriageway was built. One Launceston man remembers returning from business at Pentewan and being concerned to see that the toll-house at Lockengate had been demolished. Concern later turned to dismay to see that Palmers Bridge toll-house had also been torn down.

Later, he tackled Mr Leslie Hooper, county councillor for the area at the time, about the sudden demolition of the toll-house. The man was told in no uncertain terms that the demolition had to be done quickly and quietly so that no darned fool like him would have the chance to get a preservation order slapped on the building; the whole road scheme, now taking the dual carriageway over long stretches of the moor, would have been thrown into jeopardy! Until a few years ago it was possible, in season, to see snowdrops, daffodils and lilacs growing in the old toll-house keeper's garden! In recent years most of it has been scrubbed out, and nothing tangible remains.

Flowing on the river neatly avoids Halvana fir plantation, which was established between 1932 and 1958. For all its vastness or maybe because of it, Halvana has not been a stranger to controversy. In June 1992 it was suddenly announced that Exchem PLC had come to an

Once a toll-house stood at Palmer's Bridge, a mile or so from Jamaica Inn. In the mid-nineteenth century, Richard and Anne Barber, relatives of a local lady, were toll-keepers at this isolated spot. The toll-house was demolished in March 1986 to make way for a road improvement scheme.

arrangement with the Forestry Commission to seek permission from the North Cornwall Council to build an explosives store in Halvana plantation. A new access road would be built and the vehicles would not pass anywhere near existing houses, though the Forestry Commission revealed that permission had been granted for a new road leading directly off the A30 trunk road into the Halvana and that there had been access problems in the past and the new road should relieve the problem of traffic to the site passing too near local homes. Understandably there was a tsunami of protest from many local residents. A protest petition was quickly raised, the Liberal MP for North Cornwall, Paul Tyler (now Lord Tyler of Linkinhorne) became involved and wrote letters, explaining that the site was unsuitable on access, environmental and recreational and safety points of view, to John Gummer, then the Agricultural Minister, and spoke out strongly against the plan, as did local councillors Ken White and Leslie Hooper. At one time over seventy concerned local residents packed a public meeting to air their views. Numerous points of objection were raised including access to the site and the fact that Halvana plantation is a public amenity with designated walks. One lady said that the vehicles carrying the explosives would be passing within a few feet of her bedroom window! Another said that the roads were narrow, with high hedges and not enough room for a lorry to pass a pedestrian, and that they were totally unsuitable to take any increase in the volume of traffic generated by the store. Others feared that the site would be a target for terrorists and other criminals. A spokesman for Exchem PLC explained that they had been given notice to vacate their site at Bovey Tracey and that the site at Halvana, though unmanned, would be secure and that electronic devices would be installed.

Less than a mile away is the famous Jamaica Inn, one of the most iconic buildings on Bodmin Moor. The inn is known to millions of people worldwide since Daphne Du Maurier made it the central location of her famous novel, *Jamaica Inn*. Jamaica Inn has a colourful history; it was built in 1754, and at that time was known in the tithe apportionment as the

New Inn and was a favourite watering hole for local farmers and villagers. In the eighteenth century, Jamaica Inn was a regular stop for coaches where the horses were changed on the arduous route from Launceston to Bodmin. When a member of the Rodd family who had held the post of Governor of Jamaica retired and returned to his family home at North Hill, the inn then took on its exotic-sounding name, Jamaica Inn, in his honour. It is believed that the famous Jamaica Inn sign is probably the most photographed inn sign in the country, if not the world.

Old photographs of Jamaica Inn show that adjoining the inn there were stables and barns which have been converted into a restaurant, gift shop and the Daphne Du Maurier museum. At one time there was a public footpath which ran straight through the inn's beer and wine cellar and across the moor to Rough Tor and Brown Willy, which over the years caused no end of problems. It is reported on in the *Launceston Weekly News* of June 1976, when the dispute had come to a head. Apparently, the owner and the barman, Anthony John Watts and John Edward Lee, had been taken to court at Launceston, accused of wilfully blocking an age-old right of way which had been obstructed by the erection of a barrel store. The obstruction was challenged in the court by villagers and visitors who were having none of it and they won the case. The two men challenged their convictions and the case went before the Crown Court at Bodmin. Even though the owner claimed that the right of way skirted round the end of his property and that he had made a well-defined footpath for the people to use, the judge, Mr C. M. Lavington, after hearing and considering all the relevant evidence, dismissed the appeal

Jamaica Inn was once a temperance house but is now a fully fledged inn. In earlier centuries it was an important staging post for horse-drawn mail coaches. More recently Jamaica Inn was owned by Alistair Maclean, the novelist. Today Jamaica Inn is a 'must-visit' stop for countless fans of Dame Daphne Du Maurier.

by the two men. Today the original footpath sign is currently on display. Ironically, fifteen years later, the footpath was disbanded when in August 1991 a new section of trunk road over Bodmin Moor was opened, so the footpath no longer exists on inn property.

Every Boxing Day morning the East Cornwall foxhounds meet at Jamaica Inn, when the cobbled courtyard fronting the inn is jammed by a large mounted field and several hundred footsoldiers crowd the car-park and roadway as a show of strength and support for the pack.

Jamaica Inn is synonymous with Daphne Du Maurier, who chose the inn as the setting for her famous novel, which was first published in 1935. Apparently the idea for the novel came about after Daphne Du Maurier and her friend Foy Quiller Couch, daughter of the celebrated novelist, became lost while riding on the moor and had a chance meeting with the then vicar of Altarnun. It is easy to imagine the foreboding which the heroine of the story, the orphaned Mary Yelland, felt when the coach dropped her outside the inn and drove away, leading to a less-than-warm welcome from her uncle, Joss Merlin. Since the spectacular success of the novel, Jamaica Inn has become a mecca for Daphne Du Maurier devotees.

The Du Maurier experience was enhanced by John and Wendy Watts, who incidentally purchased Jamaica Inn from novelist Alistair McLean. After Dame Daphne's death in 1989, Mr and Mrs Watts had the idea of forming a Du Maurier museum, which was opened in 1990 by Robert Browning, Dame Daphne's grandson. The artefacts came from Dame

The meet of the East Cornwall Foxhounds at Jamaica Inn on Boxing Day 2011, which was attended by hundreds of followers.

General view of the meet and its many followers and not a protester in sight!

This picture shows a typical Bodmin Moor farmer firmly astride his trusty mount. One imagines that the rider has probably been a keen hunt supporter since he first began to ride!

Daphne's home, Kilmarth, near Par. Visitors can admire a portrait of Daphne, painted by Harington Mann when she was a teenager, with a far-away look as if she's imagining another novel. Devotees of Dame Daphne can see several pieces of her property which were bought at auction, after spirited bidding from the room and on the telephone, by John and Wendy Watts, the then owners of Jamaica Inn. Dame Daphne's Sheraton writing desk, which realised £7,500, her manual typewriter, telephone, photographs, glacier mints and a packet of Du Maurier cigarettes. There are also a number of family photographs, including one of Gerald Du Maurier, actor manager; a portrait of Dame Daphne, especially signed for Jamaica Inn; a photograph of Lt-Gen. Sir Frederick Browning, along with numerous photographs of boats and archery.

Another part of the museum is given over to smuggling in various parts of the country. There are cutlasses, naval swords, lantern barrels and much more, as well as items showing some ingenious ways in which the smugglers tried to outdo the customs men. A lady's shoe with a hollow heel, a corset with secret pockets, a turtle shell which could be hung on a wall, hollow tubes for contraband which could be hidden in a ship's mast, and a way of smuggling live birds in stockings hidden under heavy coats across country borders.

Not surprisingly, legend has it that Jamaica Inn is haunted. Apparently, so the story goes, a man was in the bar quietly supping his pint when he was suddenly called outside. He left the bar and was never seen alive again. Next morning his dead body was found in the

The Sheraton writing desk and trusty typewriter which once belonged to Daphne Du Maurier and on which she wrote her famous novels, particularly *Jamaica Inn*. These items are now prized exhibits at the Jamaica Inn Du Maurier exhibition. (Photograph courtesy Jamaica Inn Du Maurier exhibition)

courtyard and he is reputed to still haunt the inn. Nearly thirty years ago, one Launceston man had friends, a level-headed business couple, visiting from the Isle of Man. High on the list of places to visit were, strangely, Dartmoor Prison at Princetown and Jamaica Inn. It was an early autumn afternoon when the pair checked themselves into Jamaica Inn to stay the night. A convivial evening was spent chatting by the open-fireplace. Apparently the lady in the party was awakened in the small hours of the night by the sound of horses' hooves in the courtyard. On looking out of the window she saw a man dressed in old-fashioned clothes carrying what she described as a small barrel into the building. Her husband teased her, saying she must have dined too well at the inn during the evening. The lady was adamant about what she had seen in those moonlit hours and stuck to her story until the day she died. But who knows! In January 2004, Sky Television's *Most Haunted* team came to film at Jamaica Inn. Interestingly a psychic, Derek Acorah, in a piece of film not shown during the television programme, examined a length of old chain and declared it to have a connection with some ghastly crime which took place at Jamaica Inn many years ago. The chain is on view in a lighted alcove and, perhaps not for the squeamish, can be touched by visitors! From time to time Jamaica Inn hosts haunted weekends which have proved to be very popular!

At Palmers Bridge the River Fowey suddenly dives under the road bridge and heads off through woodland towards St Luke's Bottoms.

This model exhibit shows a typical eighteenth-century smuggler among other fascinating exhibits related to smuggling which are on view at the Jamaica Inn smugglers exhibition.
(Photograph courtesy Jamaica Inn Smugglers exhibition)

2

The Peaty River

In the early part of the nineteenth century one of the big landowners hereabouts, Squire Rodd of Trebartha, had the idea to create a new parish, so he snaffled land from the neighbouring parishes of Cardinham, St Neot and Altarnun, for his 'Bold Venture', which name gradually became the Bolventor we know today.

Once there was only the world-famous Jamaica Inn and a few granite-built cottages set in the middle of the moor, where the track from Launceston to Bodmin crossed the bleak moor. There was also the village school and the parish church. At one time there were two small shops in the Jamaica Cottages. One was run was by Mrs Emma Nottle, who sold groceries, tobacco and stationery. By all accounts, Mrs Nottle started her shop with all the money she had amounting to 10s (50p), and she successfully ran the shop for eighteen years. Almost next door, a Mr and Mrs Couch successfully ran a similar shop.

Bolventor National School was built in 1846 and did excellent service until the board school opened in 1878; it was later known as the County Primary School. Later the old school became the village reading room. At one time the school had around 120 pupils, but since senior pupils now go off to the towns the number has dwindled. Incidentally, the *Cornish Guardian* reports that the BBC broadcast a programme from the school in October 1948 which was repeated the following April. During the programme the children sang carols, gave recitations and performed two plays. During the school Christmas party the children were given a surprise! Apparently the BBC officials were so impressed by the performances of the children that, as a thank-you to them, they gave Mr Bennetts, the headmaster, money to buy each pupil a bag of sweets and later to be taken to a film show. In July 1990 a parents group was formed to fight the closure, and at one period there was a demonstration of parents outside the building. Later the Cornwall County Council, at a meeting at County Hall in Truro, decided that the school pupils should be sent to the school at Five Lanes in Altarnun on its closure. The school finally closed in 1992 after 114 years; young Tina Hooper was the last pupil.

Holy Trinity church was built by Squire Rodd to serve the new parish of Bolventor. Around that time reliable forecasts declared that Bolventor would see a marked rise in population growth, but they were wrong as the population growth failed to materialise. The foundation stone was laid by Master Francis Rashleigh Rodd of Trebartha Hall in July 1846. The ceremony was witnessed by several local clergy, including Reverend R. H. Tripp of Altarnon, Reverend C. Rodd of North Hill, Reverend R. S. Stevens from South Petherwin, Reverend H. Grylls from St Neot and a large number of people from the surrounding area.

After the stone was laid, Reverend Tripp said prayers, which were followed by communal singing. The church was the centre for a population of around 250 people, all living within 2 miles; more lived between 4 and 6 miles distant and others more than 6 miles away from the churches at Altarnon or St Neot. The nave was 41 feet by 14 feet, the chancel 17 feet by 10 feet, the two transepts each 13 feet by 13 feet, and there was seating for 150 persons. The cost was £666, and the Diocesan Church Building Society gave £100. Mr Rodd generously promised to endow the church with £45 a year from Dry Works farm and an additional £5 per year for repairs and to build a parsonage house. The whole project was to be completed within a year.

The church, which was to be served by the vicars of Altarnon, Cardinham and St Neot, was consecrated in July 1848 by the Bishop of Exeter (there was no diocese or Bishop of Truro at that time). From the registers it appears that after consecration the early services were carried out by Reverend R. H. Tripp, vicar of Altarnon. Reverend Tripp performed the first baptism, that of Fanny, daughter of James and Mary Weale of Palmers Bridge, toll-house gatekeepers. The first incumbent of Bolventor, appointed by Squire Rodd, was Reverend William Avery, who must have found his first funeral service quite poignant, as it was that of three-year-old John Bone of Buttern Tor. The first marriage on 18 June 1849 was solemnised between John Herring Dawe of Deep Hatches and Ann Lethlean of Pridacombe. Apparently neither could sign their names so both made their X mark.

Bolventor church has a fascinating history. It was built at the cost of £666 and was completed within the year of the foundation stone being laid. At the time of writing (August 2012) Holy Trinity is undergoing roof repairs and being converted into a dwelling. (Photograph Colin Barratt collection)

Another marriage, solemnised by the Reverend W. A. Kneebone of Altarnon, caused great interest and was reported on in the *Cornish Guardian* of June 1939; that between two older people, Mrs Emma Nottle, aged seventy-eight, and Mr Joseph Giles, aged eighty-three years. Apparently Mrs Nottle was married to her former husband for thirty-six years and had been a widow for over twenty years. Mr Giles was a native of Lerryn who had been living at Dobwalls. The couple apparently met when Mr Giles came into Mrs Nottle's shop to buy some tobacco on a day when it was pouring with rain. The good lady kindly offered him a cup of tea, which was gratefully accepted, and three months later they were married. No wedding day goes without an anxious moment or two and this day, so the newspaper report tells, was no exception. Apparently the best man failed to turn up, so a local newspaper reporter quickly stepped into the breach so that the ceremony could proceed, and afterwards he acted as bell-ringer. The church was crammed with well-wishers; many were forced to stand in the aisles, and a huge crowd waited outside for the newly-weds to appear to shower them with rice and confetti.

In August 1965, so the *Cornish Guardian* reports, Holy Trinity church at Bolventor was re-hallowed by the Bishop of Truro, Right Reverend J. M. Key, after a £1,750 restoration programme. Mr W. P. Drury, architect at Exeter Cathedral, oversaw the work. During March 1985, the *Launceston Weekly News* reported that closure loomed large for Holy Trinity due to lack of congregations. The Church Commissioners in London were actively considering its closure and, if agreed, unless there were calls to save it, it would be the first Anglican church in Cornwall to be declared redundant. If closed, the church would be demolished as it is difficult to redevelop as it stands in the middle of the cemetery. The last regular services were held at Holy Trinity in the early 1980s. At one period, in a bid to save Holy Trinity church, Mr Peter Halls, a local farmer, tried to negotiate between the Church of England officials and the Methodist Church officials at Liskeard after it was suggested that the building be taken over by that community, but to no avail. During talks with what was then North Cornwall District Council, an idea was floated that Holy Trinity should be converted into a crematorium, which suggestion, as might well be expected, did not find favour in the locality. Support for Holy Trinity came from Dr Michael Ramsey, Archbishop of Canterbury, who was a relative of a former incumbent of Holy Trinity. He also used to holiday in the area and consequently knew the church well. At the time of writing (September 2011) the church has been sold and re-development work has begun.

In January 1952 Bolventor was 'on the air' in a radio programme called *The Years Round at Bolventor*, broadcast by the West of England Home Service. Listeners heard schoolmaster Reg Bennetts, president of the Women's Institute Mrs Bray, Mary Parkyn, Reverend Kneebone selling harvest festival produce in aid of the church, with spirited and friendly rivalry in the bidding between Sam Sleep and Eddie Truscott and Maurice Boney, all well-loved local characters.

A mile or so away is the legendary Dozmary Pool, one of the strangest and most haunted places on Bodmin Moor. This high-altitude (for Cornwall) round sheet of water (690 feet above sea-level) is a mile or so in circumference. It rests under a vast sky, steeped in mystery and legend and even on the sunniest of days, with only the plaintive cries of the birds, exudes an eerie atmosphere all of its own. It is little wonder that the ancient Celts believed that somewhere in the midst of Dozmary Pool a secret entrance led to the underworld! It is believed that in the time of Dungarth, the Cornish king hunted deer on the moor, and that a moated lake dwelling stood either in the middle or on the edge of the pool. At one time a pile of stones was discovered at the lake's edge, which served to give the ring of truth to the belief. Arrowheads and worked flints have been found around the lake.

Over the decades several eminent travellers and historians have visited Dozmary Pool. John Leland, King Henry VIII's antiquary, who came to the lake in 1509, later described it, writing that 'Dosmery Pool is of a length of eleven arrow shots and of breadth one and standeth on a hill'. Other travellers who found their way here were Norden, Carew, Celia Fiennes and the eighteenth-century poet laureate Tennyson, who afterwards wrote about King Arthur, Excalibur and the Lady of the Lake, which epic drew Dozmary Pool to the attention of travellers and put it on the map. Some historians have likened Dozmary Pool to the Dead Sea due to its almost lifeless appearance. In earlier centuries, legend had it that Dozmary Pool was bottomless and it was taken as gospel. A story to back up this belief tells how a faggot of wood was thrown into the pool and was miraculously washed up some miles away at Fowey. This supposition was disproved during the exceptionally dry summer of 1896 when the pool dried up, and it was seen that it is actually fed from the bottom by a number of small springs. Another story concerns Jan Tregeagle, the wayward steward to Lord Robartes at Lanhydrock, who because of his misdemeanours was sentenced to dip Dozmary Pool dry with a leaky limpet shell. On stormy nights his angry roars at the hopelessness of the task can be heard on the moor as a screaming Tregeagle, being chased over the moor by a pack of headless hell-hounds, makes his way to Roche Rock. Legend has it that he put his head through the chapel window and so gained sanctuary.

Steeped in legend, the wild expanse of Dozmary Pool exudes an eerie atmosphere. Once there was an ice house at the edge of the lake. Ancient flints and arrowheads have been found around the lake.

It was after the Battle of Camlann that Sir Bedivere was commanded by King Arthur to throw his sword, Excalibur, into the lake. Sir Bedivere must have been amazed when, as the sword arced out over the lake and was just about to hit the water, a silk-swathed arm broke the surface, the hand grasped the sword hilt and it was carried into the murky water of the lake.

No one knows exactly when Dozmary Pool became a favourite venue for Sunday school treats, but one special occasion occurred in August 1880. By all accounts Mr William Prout landed a new job as district surveyor of highways, which necessitated a move for himself and his family to Holsworthy. Apparently the teachers and pupils of the local Wesleyan Sunday school decided to hold a picnic at Dosmere (*sic*) Pool on Altarnun Moors. Mr Prout must have been held in high regard as other local chapel Sunday schools, principally from Trebullett, North Hill, Bathpool, Polyphant and Congdon Shop, were all invited to join in. On the great day everyone gathered at Plusha. The North Hill brass band led the lengthy procession of around forty wagons, conveying around 600 people as well as all the equipment needed to provide a grand tea. The caravan halted at Five Lanes, where the band played for a short time before resuming their places at the front. All the wagons were left at Bolventor and everyone walked the mile or so to Dozmary Pool. At the poolside there was boating, cricket and food enough for everyone to enjoy two sittings! At Plusha on the return journey, three hearty cheers were given for Mr John Prideaux, who organised the event, and three more for the farmers who lent the wagons.

At one time there was an ice works at Dozmary Pool. The ice house consisted of a long rectangular enclosure, the floor of which was paved with large granite slabs, cut into the hillside with an exit to the pool. The enterprise was worked by several local farmers. During the winter months the ice was cut into large blocks and floated to the edge of the pool, from where horse-power hauled the ice up a ramp to the ice house. When the enclosure was full, it was covered with turf. The ice would remain in the ice house until the summer months. Horses and carts would take the ice to Moorswater near Liskeard. At Moorswater it was transferred onto canal barges and floated to Looe. At Looe it was packed tightly round the newly landed catches of fish, ensuring their freshness during the rail journey to the London Markets and destinations all over the country. At one period during the late nineteenth century, a Mr James Henderson of Truro established a new local industry. He had been since 1877 working on a patent for the crushing of ice or snow and compressing it into blocks for commercial purposes. Henderson formed the English National Ice Co., which would only use natural ice. He had a depot for ice at Sourton near Okehampton and at Dozmary Pool. He also expanded his operation to include building reservoirs, storehouses, plants and machinery at other locations. Many tons of ice were sent annually to various ports, mainly for preserving fish being sent to markets from Hayle, St Ives, Mevagissey, and Plymouth. In December 1877 the *West Briton* reported that a hurricane struck the area and great damage was done to property throughout the district. At Dozmary Pool the roof of an ice works was blown off by the wind and carried a considerable distance. The ice works closed down around 1900.

During one of the harshest winters on record in the 1940s the lake froze over and one local lady, then a Miss Manse, who at the time was employed by a Launceston photographer, well remembers skating on the lake and having her photograph taken, which was later featured in the local newspaper. Legend has it that at the time a small car was also driven over the ice from one side to the other. Apparently, at one time when the lake froze over during the

Dozmary Pool, probably the most haunted sheet of water in Cornwall. Here Sir Bedivere threw King Arthur's sword, Excalibur, into the lake. Sceptics may scoff but just because Excalibur has not been found. It does not mean that it is not there!

These youngsters seem to be enjoying themselves boating on Dozmary Pool, which was once part and parcel of any celebrations connected with nearby St Luke's chapel.

Christmas period, local families gathered at the pool, ventured out onto the ice and sang carols.

It is inconceivable that at one time water shortages were an annual summer problem and a proposal was made to build a dam across the River Fowey at Lamelgate, which would have effectively drowned farms and around 700 acres of land would have been lost. The reservoir would have supplied consumers in Devon, Cornwall and Plymouth with around 40 million gallons of water daily. Thankfully the proposal never came to fruition and this stretch of the Upper Fowey Valley remains.

A mile or so from Jamaica Inn is St Luke's chapel, which has long been held in affection by generations of moorland people. The tiny new chapel's opening day in February 1858 was an eagerly awaited occasion for the local Bible Christian community. The *Cornish Times* was happy to report that on the day, Mr T. W. Garland of Tavistock preached two impressive sermons and the chapel was full again on the following Sunday when Mr Wooldridge of Camelford delivered a sermon much to the edification and profit to all those present. By all accounts, over sixty people sat down to a public tea. Around £6 was raised during the three services and around £35 was given in donations, which was considered noble and praiseworthy to the parishioners. Over the years St Luke's chapel congregations consisted mainly of local farming families who sometimes walked long distances across the moor in the summer and in the bleak winter months to attend services.

Here it would appear that there is something of a mystery; the opening date on the slate plaque reveals a discrepancy of some thirty-three years. Now all can be revealed. According to the *Royal Cornwall Gazette*, due to the inadequate size of the Sunday school, the small chapel had apparently been unable to carry on its work. It was decided to convert the original chapel to a Sunday school and to build a new chapel, hence the plaque date of 1891. Mr W. H. Olliver (*sic*) of St Cleer was asked to build the new chapel. It is reported that he provided a substantial, neat and commodious chapel to sit around 140 persons. The honour of laying the foundation stone went to Mr H. Smith of Nuneaton and was witnessed by a large crowd. During the proceedings Mr Spillett, superintendent of the Liskeard circuit, said they had made a good start. The debt of the old premises had been wiped out and there was a good balance in hand. He did not think that Methodism had completed its mission yet. These remarks were met with hearty applause.

Well within living memory, Dozmary Pool, a mile in circumference, was the venue for St Luke's chapel Sunday school tea treats services, attended by hundreds of people, who made it an annual pilgrimage. Boating was another feature of the day. A boat was housed in a small building by the lake. A few days prior to the anniversary celebrations on a Sunday in June, the boat was taken to the lake, floated onto the water, filled with stones and sunk. This was not an act of wanton vandalism designed to scupper the celebrations, as it might at first appear, but the time-honoured way of making the sure that by tea treat day, the boat boards would have swollen enough to ensure that the vessel would be watertight. Food, tea urns and seating were all taken to the site by horse-and-cart and in later years, tractor and trailer.

In October 1965 after an extensive renovation programme the chapel was re-opened. At the ceremony the chapel door was unlocked by Mr W. Turner, who at the age of eighty-five years was the oldest member. The chapel was packed to capacity for the service, which was

St Luke's chapel at Bolventor, one of the smallest chapels in the district, has sadly closed.

conducted by the Rev R. C. W. Smith of Liskeard, and the special preacher was Reverend R. James of Bude. There were also soloists and an organ recital. The *Launceston Weekly News* reports that, during the proceedings, the electric light was switched on by Mr A. Stephens and the new decoration could be seen in all its glory. There was a concert in the evening at which a choir sang and a local group of handbell-ringers performed, followed by a supper. Sadly St Luke's closed in November 2007 and at the time of writing (April 2012) the building has been sold and is being converted into a private dwelling.

Below St Luke's, in what the locals fondly call 'St Luke's Bottoms', the river is edged by overhanging willow trees and in the distance is the great mass of the Smallacombe Downs plantation. From below Bolventor the River Fowey flows close to the roadway for a distance of over 6 miles without a junction. Interestingly, this stretch of highway is believed to be one of only a handful of such stretches in the country. The River Fowey flows peaceably for the most part through the valley, though in winter the river makes itself felt as the roadway is liable to flooding. The highway continue to progress in tandem with the river to Higher Harrowbridge Saddlery.

Jane Talbot Smith is a master saddler, the first in Cornwall, whose two Hungarian Vizsla dogs greet visitors with unbridled canine affection. Mrs Talbot Smith has always been connected with horses. She spent nearly three years building horse-drawn vehicles and then, after falling in love with leather, decided to study rural saddlery at Cordwainer's College, Hackney, London. Later, to further her career Jane worked as an improver with Bridleways of Guildford under the guidance of master saddler Les Coker, where she undertook every facet

A day of great celebration. St Luke's chapel re-opening in October 1965. (Photograph courtesy Cornwall Centre, Redruth)

This picture shows the bridge at the entrance to Wimalford farm. Unfortunately, due to many years of weathering it is now impossible to read the granite plaque.

of harness work and saddlery. The highest accolade came to Jane in 1985 when her work was expertly assessed and she was accepted into the eminent Society of Master Saddlers. Inside her workshop, where visitors are always welcome, Jane Talbot Smith practices her craft with her beloved dogs in close attendance. Only English leather is used and it is clearly evident that everything is made entirely by hand, using a range of specialist tools. To make a saddle from start to finish takes around thirty hours of concentrated work. Over the years the range of leather work has expanded to include belts, bellows, cartridge cases and dog collars, which items go all over the world. Jane Talbot Smith, having hailed from St Veep further down the river before moving away, describes herself as being like a salmon which is now happy to have returned to its own river!

Beside the roadway, the river hurries toward Carkeet. At Carkeet are the remains of the only brickworks on Bodmin Moor. By all accounts it was only in operation for a short period between the late nineteenth century and the early decades of the twentieth century. One of the kilns and its chimney stack, with its diamond patterning, still stands proud in the landscape. The bricks are reddish in colour and marked 'Liskeard' and a few are scattered around the site. Further downstream is Trekieve School, where Miss Mary Marrack, a well-

Above left: This picture shows master saddler Jane Talbot Smith hard at work in her workshop at the Higher Harrowbridge Saddlery.

Above right: Some of the wide variety of leather items, including saddles, horse harnesses, dog leads and collars, made by Jane Talbot Smith in her workshop at Harrowbridge Saddlery.

known local character from Commonmoor, told me she once taught. The river flows close to the old school building of 1876 and Miss Marrack recalled how in the winter months the river was always liable to flood, and to ensure that her pupils could get home safely she had a tried and trusted contingency plan! A local farmer used to convey the pupils safely through the flood water on his tractor. Trekieve School closed in 1944.

Near here too is Sibley Back reservoir and recreation centre. The dam and reservoir was a six-year project by the East Cornwall Water Board. The dam itself is 750 feet long and 70 feet high, the maximum depth of the water is 80 feet and the surface area of the reservoir covers 140 acres, holding 700 million gallons. The cost of the scheme was £685,000. Water from Sibley Back can be sent to Bastreet near Launceston and also to Polperro, Looe and the Torpoint areas. A minimum flow of some 3 million gallons a day can be released into the River Fowey. On opening day in June 1969 a short dedicatory service was conducted by the Right Reverend Maurice Key, Bishop of Truro, and the dam was opened by Sir John Carew Pole, Lord Lieutenant of Cornwall, who unveiled a commemorative plaque and switched on the pumps, which can if necessary pump water to other areas. Invited guests were brought to Sibley Back from Liskeard by a fleet of coaches. The general public were permitted to watch the ceremony from the top of the dam.

Sibley Back quickly grew into and remains a popular leisure amenity for East Cornwall. Apparently not everyone was happy. According to the *Cornish Guardian* of January 1971, scarcely eighteen months after the opening ceremony, concern was expressed about the marked exodus of wildlife at Sibley Back due to canoeing. According to local naturalists, wildlife had seemingly been 'almost decimated'. The observation was mentioned by Mr Leonard Mitchell at a meeting of the Liskeard Rural Council. It had been suggested that a string of buoys should be strung across the lake near Sibley Back marsh. The rural council

Remains of one of the buildings at Carkeet brickworks, the only brickwork in the Fowey Valley.

The chimney at Carkeet brickworks with its distinctive diamond-shaped pattern is a gentle reminder of what was once a thriving concern.

agreed to ask the East Cornwall Water Board to consider the point. Today, among the facilities offered are boating, sailing, canoeing, windsurfing, birdwatching and fishing, the lake having been stocked with thousands of brown and rainbow trout.

On Midsummer's Eve an interesting ceremony, currently organised by Liskeard Old Cornwall Society, takes place on the lakeside. Dating originally from pagan times, a bonfire, one of a chain stretching the whole length of Cornwall, is lit on St John's Eve, to welcome in the summer as the sun is at its highest on St John's Day. Seemingly the custom died out in the nineteenth century but was revived in 1929 by the Old Cornwall movement. A local worthy is invited to light the bonfire and prayers are offered by a clergyman. The 'Lady of the Flowers', usually a young lady, is invited to cast a bouquet of wild flowers and herbs which have good, medicinal and bad influences, bound with white, blue, red, green and yellow ribbons, into the flames. After the ceremony there is traditional Cornish music, singing, dancing and pasties!

It was here, near Sibley Back, that I met Joan Bettison, a lady who was born at Darite on the edge of the moor and spent most of her life on the moor and consequently had a vast amount of local knowledge about the area and its folklore. She also took great pride in the fact that she was the last 'postie' in the Upper Fowey Valley who delivered for Royal Mail on foot or with her bicycle! It was while I was driving Joan to Commonmoor that she noticed the wart which I had on my hand for several years. Though nothing more was mentioned about the wart, sometime later it mysteriously disappeared. It was some considerable time later that I discovered that Joan was a local charmer and she had almost certainly charmed it away.

Members of Liskeard and Launceston Old Cornwall Societies and their friends enjoying the Midsummer's Eve bonfire celebrations at Sibley Back reservoir leisure centre in June 2011.

Above left: Sibley Back reservoir is a popular centre for boating, canoeing, fishing and windsurfing. In this picture, a wind surfer is in full sail.

Above right: Below Dreynes Bridge, the River Fowey cascades through Golitha Falls, one of the most spectacular stretches on the upper reaches of the river.

Charles Henderson tells that the earliest bridge over the Fowey at Dreynes dates from 1362. The present bridge, dating from 1876, is similar in style to Ashford Bridge, further downstream. At one time there was a Mission Church at Dreynes. There is a pleasant walk under the beech trees beside the river and a nature reserve is nearby.

Leaving Trekieve Bridge, the River Fowey abruptly changes direction and heads toward Dreynes Bridge, where there is a nature reserve. The earliest mention of Dreynes Bridge comes in 1362, then again by Leland in 1535. The present single-file, four-holed bridge of heavy granite moorstone was built in 1876. It was here that I happened to meet Mr Kent Grossmeyer, an American who has been living in nearby St Neot for around ten years. When I mentioned his American accent, he told me laughingly that it will not go away, despite eating countless Cornish pasties. Mr Grossmeyer has clearly explored the upper reaches of the River Fowey and ranks it as 'a great little river' and he intends to follow its course down to Fowey.

Flowing under Dreynes Bridge, the Fowey widens as it glides under an avenue of beech trees, considered by some to be one of the most picturesque stretches on the upper reaches of the river. In spring the pathway is edged with anemones and bluebells, in summer the woodland is dappled with shafts of sunlight; later it resembles a great cathedral of coloured autumnal leaves and in winter, tinged with frost, if decorated with snow they look like the skeletal remains of prehistoric beasts.

Downstream is Golitha Falls, pronounced (Gol-ee-tha). When the river is running high the sound of the falls can be heard long before they come into view. At one time, fish making their way upstream to spawn found their way blocked by huge boulders. Some of these were blasted away but a series of spectacular falls still remain.

3
King Doniert Country

Beyond Dreynes Bridge and Golitha Falls, the River Fowey turns to caress St Cleer, St Neot, Warleggan and Cardinham, a clutch of fascinating parishes, each with a tale to tell.

Not far from Dreynes Bridge, two curious granite stones stand by the roadside leading from Redgate to Minions. This is King Doniert's or Dungarth's stone, or the Half Stone and the other Half Stone, dating from the Dark Ages. The stones are carved with interlaced Celtic designs and the inscription 'DONIERT ROGAVIT PRO ANIMA' – 'Doniert asks prayers for his soul'. Doniert is thought to be one of the last Cornish kings, the Cornish having been routed by the Saxons. Doniert was drowned in the River Fowey above Golitha Falls in AD 878. The stone was rescued and later erected in its present position by Liskeard Old Cornwall Society. At the same time a protecting wall was built around the site. In spring the site is dashed with daffodils.

St Cleer parish, extending over 10,000 acres, lies on the southern side of Bodmin Moor and stretches up from the River Fowey. The main settlement is St Cleer village, though there are several smaller communities; Tremar, Lower Tremar, Tremar Coombe, Darite, once known as Railway Terrace, and Crows Nest. Since the early nineteenth century, when St Cleer was the centre of the copper mining boom, the general population of the village has risen steadily. The upward trend continues, as new roads make for easier commuter access to Plymouth.

Hocking's House Chapel, now converted into a private dwelling, stands on the edge of St Cleer village. It was not the first chapel in the parish, as that singular honour goes to Redgate and there is a connection between the two places of worship. On 20 March 1939 a special ceremony took place at Hocking's House Chapel, St Cleer. The original stone of the old Redgate Methodist Chapel, dated 1812, which had been on the property belonging to Mrs Warren, who was living in America, was placed on the chapel at Hocking's House. The wording on the stone, which was uveiled by the Rt Honourable Isaac Foot, read: 'This granite quoin is the date stone of the first Methodist chapel in this parish, erected at Redgate 1812.' The explanatory carving was the craftsmanship of Mr C. Carne of Tremar Coombe and much of the groundwork involved with the project has been undertaken by Mr T. J. Lang of Trevorrick and a band of volunteer helpers.

St Cleer church tower is one of the tallest in the Cornwall. Scholars generally agree that there was a wooden daub-and-wattle church on the site as early as the eighth century. In the fourteenth century the church was enlarged and the north aisle added, and parts of these foundations can be seen in the churchyard. The present church dates mostly from the

King Doniert's inscribed stone near St Cleer. King Doniert was drowned in the River Fowey, near here in AD 875. The stones are decorated in the ninth-century style. Liskeard Old Cornwall Society laid out the site.

Hocking's House chapel on the edge of St Cleer village has been recently (2012) undergone extensive alterations and is now a private home.

fifteenth century, though traces of the Norman church remain. Extensive restorations have taken place over the decades. Inside the church there are a series of seventeenth-century text boards which are unique in Cornwall. Sadly, several old bench-ends were ripped out and burnt. In 1850 the restorers really went to town: windows from an early period were replaced and the north wall was rebuilt. The clock was the gift of a Mr Glencross, who also donated the glass in the east window of the chancel in memory of his family.

Near the church is the sixteenth-century holy well, which reputedly has a cure for madness. In his comprehensive book on Cornish holy wells, the Reverend Lane Davies records that in 1858 Lt Henry Rogers R.N. purchased a waste plot of land and the ruins of a holy well at St Cleer. In 1864 Captain Rogers and his relatives restored the holy well as a memorial to the Reverend John Jope, his grandfather, who was the vicar of the parish from 1776 to 1844.

During the First World War, Isaac Foot, a lawyer practising in Plymouth, lived at Ramsland. The late Wilfred Hoskin remembers him walking to Liskeard station almost every day to catch a train to Plymouth. On the edge of the village there is a large rock where a plaque bears his name.

The St Neot River – or the Loveny, a picturesque name which sadly seems to have been dropped except in the name of a popular male voice choir – comes hurtling down from Bolventor and passes through St Neot to join the River Fowey. St Neot village, voted Best Kept Village in 1987 and again in 1997 – as plaques from the Council for the Protection of England on the Institute attest – is best seen from Goonzion Downs, with the church as the focal point of the scene.

St Neot church is in the top ten of Cornish churches. Legend has it that the church was built by St Neot. Scant regard is paid to the fact that St Neot lived hundreds of years before the present church was built. Tradition has it that the stone was brought to the site from a local quarry at night on a cart drawn by a reindeer. The best exterior features are on the south side and date mainly from the fifteenth century. The north was built a century or so later and is against the steeply rising hillside. Its fifteen stained glass windows dating from the early sixteenth century are remarkable. In the middle of the nineteenth century some of the windows were showing marked signs of their antiquity and the Reverend John Hedgeland, the incumbent at the time, had them restored; fortunately, some of the original glass remains.

There are some interesting slate memorials; one to William Bere and his wife, and another giving details of donations to the St Neot poor. A fine carved wood-and-glass screen divides the belfry from the body of the church. It remembers Minnie Christini Cowan, and was erected in her memory by her brother W. R. S. Majendii in July 1910. The glass screen is etched with wheat stalks, wild flowers and birds; the wooden panels below show church-bells with the legend 'ring out the false, ring in the true'. Some roof bosses bear the date 1480. The list of vicars dates from 1266, and of interest is the Father Willis organ, built in 1884. St Neot parish is rich in old granite crosses; one rich in lace-work design stands in the churchyard and is one of the best in the county.

St Neot himself is shrouded in mystery. Some sources say that he was a special friend of King Alfred, others argue that he was the elder brother to the king and had renounced everything in preference to a hermit's life and cell. But by all accounts St Neot was a small man and when the church building was complete he discovered that he could not insert the

Stately St Cleer parish church, centrepiece of one of Cornwall's largest parishes. There are a number of hamlets in the parish. Over the centuries the church has undergone several alterations with varying results.

Inside St Cleer church there are several features of interest. There are a number of texts painted on wooden shields which were fixed to the walls in 1660 and they are believed to be unique in Cornwall. The glass in the south aisle came from a mission church at Dreynes.

Above left: St Cleer holy well, a feature of the village, is now dry and somewhat hemmed in by modern bungalow development. The late Wilfred Hoskin firmly believed that the well dried up due to what was then the Water Board undertaking work too close to the well and water being diverted away!

Above right: The late Wilfred Hoskin, one of St Cleer's real old characters. He was steeped in the history of the village. He was also a keen sportsman and played football and cricket for several local teams. In later life he played snooker and billiards.

General view of St Neot, which is arguably the most staunchly Royalist of all Cornish villages and unashamedly displays the fact from the church tower for everyone to see.

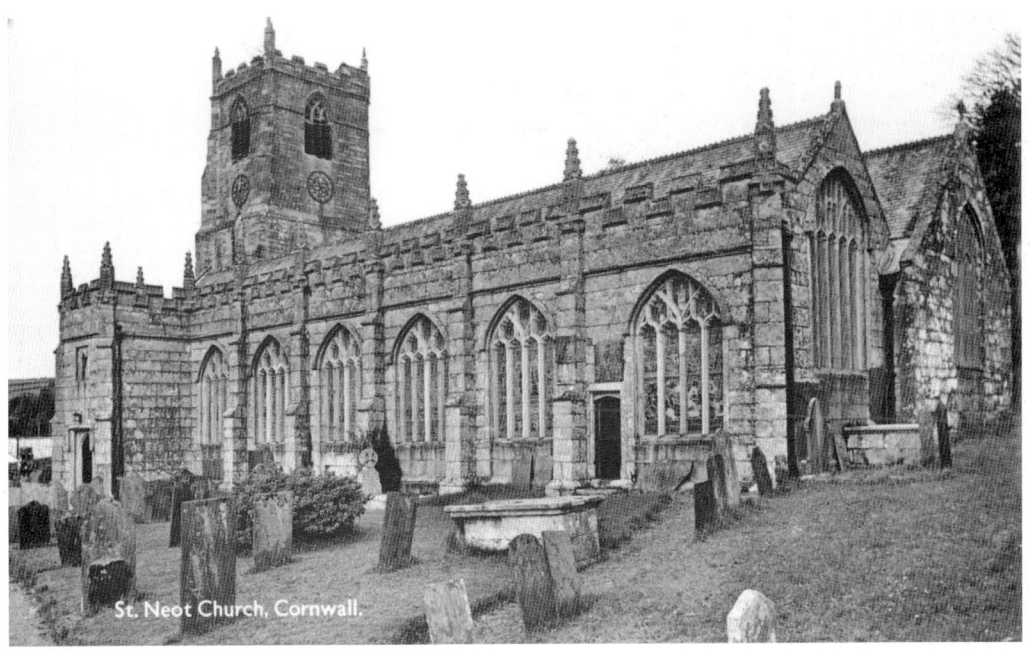

St Neot parish church and its neighbour, the popular London Inn, are centrepieces of the village. In May 2012 building is taking place to provide the village with a Rural Resource Centre, Rural Workshop, Tourist Information Point, village archive and the relocated post office and shop.

Interior of St Neot church, which is famous for its early sixteenth-century stained glass, the Creation window being the oldest.

door key into the key hole to unlock the door. However, ingenuity prevailed and a stone was placed so that he could stand on it and with uncanny accuracy he could throw the key into the hole and he was then able to unlock the door.

During the Civil War St Neot was staunchly Royalist and history tells that after the Battle of Worcester, King Charles II hid in an oak tree to escape his pursuers. At St Neot the King's miraculous escape is remembered by a custom which is little-known beyond the confines of the village. This event is remembered at St Neot as a benefactor left a sum of money so that on 29 May each year, the branch of an oak tree would be ceremoniously processed through the village to the church. The procession from the Institute to the church is led by the vicar, whose pastoral-staff is decorated with red campion. The oak branch has been provided by the Bunt family from Penlan farm for several generations and still is today. At the church the leafy oak branch is hoisted by rope up the tower – a delicate operation as the branch passes within inches of the clock-face – and is then proudly displayed from the top, where it will remain for the next year for all to see.

St Neot's holy well is a short distance from the church. Legend has it that St Neot lived on a staple diet of fish, believed to be shad, the daily supply being provided from the well. Apparently an angel told St Neot that if he only took one fish from the well then on the following morning he would find that there were still fish in the well. However when St Neot was ill his servant took two fish and cooked them. Somewhat irritated, St Neot ordered his servant to return the fish to the well, where they miraculously came back to life. The story is told in the carvings of the pulpit.

The Carnglaze Slate Caverns were once one of the best-kept secrets on this section of the River Loveny just before it joins forces with the River Fowey. Since 1973 the secret has been out. It was then that Mr Alan Pascoe, who had been born and bred by the caverns, which had been owned by his family for several generations, had the idea to open them to the public. Today thousands of people visit the caverns annually. No one knows exactly when the story of quarrying began at Carnglaze Slate Caverns, but in 1856, John Allen, in his *History of the Borough of Liskeard*, states that it is 'extremely ancient' and that in the early days there were probably no caverns. It is believed that the cavern system proper began sometime around 1780 when it was noticed, probably by a manager, that a great deal of the best slate was coming from the bottom of the steep cliff. The last company to work the quarry started operations in 1888, when a Mr Edward Pascoe was appointed as manager.

Prior to the arrival of the railway, slate was taken by horse-and-cart to St Winnow quay on the Fowey or to the small ports of Looe and Polperro, where it was loaded onto sailing vessels ready for shipment. More recently carts took the slate to the railway station at Doublebois. Later the cost of transporting the slate away from the caverns came into the equation and the operation became uneconomic and quarrying ceased. In the years before the Second World War, thousands of tons of stone were removed from inside the caverns for use in the building industry and also for roads. During the Second World War the caverns were taken over by the Royal Navy, who utilised them as a rum store. At one time there were several millions of pounds worth of 'grog', which came from as far away as Australia and South Africa, stored here. It was distributed through the Devonport Victualling Yard at Plymouth. Soon after the quarry closed, vast numbers of bats 'squatted' in the caverns but

On 29 May each year the age-old custom of raising an oak tree branch to the top of the church tower takes place. In this picture from May 2012, the Reverend Andrew Balfour leads the procession with all due solemnity as the oak branch is carried to the church.

This picture shows that the oak branch has been successfully hoisted aloft, where it will remain peering out from the church tower for the next year, when the ceremony will again take place.

The famous Carnglaze slate caverns at St Neot, near where the River Loveny joins the River Fowey. High-quality slate was quarried by hand over many centuries. During the Second World War vast quantities of Royal Navy rum was stored in its depths. Today the caverns are a popular all-year tourist attraction and venue for weddings and choirs and orchestral concerts.

seemingly they could not stand the regular disturbance nor the strong smell of rum, and they suddenly took wing to who knows where, never again to return!

The upper chamber is around 300 feet long, around 50 feet wide, with a head height varying between 10 and 20 feet, and remarkably the roof is entirely unsupported. A long flight of steps cut from the natural slate guides visitors to the floor of the caverns, barely lit, which provides the atmosphere and conditions in which the 'old boys' here would have worked. It is said that young boys came to work at the caverns and spent all their working lives here, like troglodytes in the half-light. Although the work was arduous and the hours long, there were perhaps some advantages. The caverns have a constant temperature of 52 degrees, and they were not affected by any seasonal weather conditions, and although they were working deep underground, it was seemingly a safe environment. There has never been any kind of accident at the caverns. The pool is about thirty feet deep and a clear blue, caused by millions of particles of slate dust which, over the decades, covered the floor of the caverns, and reflect their light up through the water. An eye-catching feature, relics of the working quarry, are the eye-bolts and chain suspended from the roof over the pool, part of a block-and-tackle system which was in daily use when the quarry was operational. The Carnglaze experience includes a display of minerals found in the district and a selection of old quarrying tools. Although Carnglaze Quarry ceased working over 100 years ago, it is enjoying a unique renaissance as a popular venue for choir concerts, operatic recitals, orchestral concerts and weddings.

From here the St Neot (or River Loveny) scurries on to Two Waters Foot, where it joins the River Fowey. Of all the parishes touched by the River Fowey, it is Warleggan, bounded on the southern side by the river and sliced by the Rivers Dewey and Bedalder, which is most wrapped in legend, myth, mystery and intrigue. Warleggan, originally spelt Worleggan, meaning 'High Place', is sparsely populated and is, according to Nikolaus Pevsner, one of the loneliest parishes on Bodmin Moor. The long narrow parish, a smidgen less than 3,000 acres, is seemingly blocked at either end by Temple and Braddock and sandwiched lengthways between its two larger neighbours, Cardinham and St Neot.

Warleggan village consists of a handful of houses, a chapel, the church and the former rectory, which is appropriately called the Rookery and is now a private residence. Once, tin and copper mining and granite quarrying took place at Grazeland and Treveddoe.

The parish church is built of moorstone and dedicated to St Bartholomew, references to it first appearing in 1434. It seems that in 1480 the building was extended. The granite pillars have worn carvings. The Devil's Door, through which he was driven out of the church during baptisms, has been blocked up. According to the vicar, in 1727, much of the furniture was in good order. A slate plaque in the tower tells that it was 'repeared' (*sic*) in 1754. It gives the names of John Toll and Richard Pearse, churchwardens. The steeple suffered a lightning strike in March 1818 and, according to the *West Briton*, there was a tremendous thunderstorm and the tower was struck by lightning and with a terrible crack split from top to bottom. One part fell into the church and destroyed most of the pews. An elderly clerk who had been inside the church preparing for the following day's service evidently had a narrow escape. While he was standing in the churchyard a stone weighing around 6 hundredweight struck the ground within 2 feet of where he stood. The report ends by saying that if he had remained in the church for a few minutes more then he would have been buried in the ruins. The steeple was never rebuilt, and today the squat tower is plagued with damp and green mould. Most of the furniture and original glass was destroyed in the fire of 1818. Some of the memorials are interesting; one in slate remembers Richard Bere, who died in 1618, and another depicts the royal coat of arms of 1664 on the north wall, marking the Cornish support for the Royalist cause during the Civil War.

One of Warleggan's fourteenth-century incumbents, John de Tremur, was excused from living at Warleggan on the grounds that a predecessor had allowed the rectory to fall into a state of disrepair. Ralph De Tremur was also an absentee vicar who studied at Oxford by permission of the bishop. Among his accomplishments was the fact that he spoke several languages, among them French, Cornish and Latin. He resigned the living in 1334 but later returned to Warleggan, where he robbed his successor and tried to burn down the rectory.

Over eighty years ago, another Warleggan incumbent was to make his parish known all over the country. Even today it is difficult to separate, with any certainty, fact from fiction. The story revolves around the Reverend Frederick William Densham, a bachelor in his early sixties who came to the parish in 1931, and was the last resident incumbent. Little or nothing was known about Densham before he came to Warleggan. However, he was born at South Petherton in Somerset, where his father was a Congregational minister. Densham gained his B.A. from the University of London, trained for the priesthood, was ordained in 1910 and became priest at Orford in Suffolk in 1911. At one time he travelled in South Africa and India, and he was keen on the beliefs and ideas of Mohandas (known as Mahatma) Gandhi. Seemingly Densham had little concept of life in a isolated country district and he must have found the winter months particularly trying. Prior to Densham's arrival in Warleggan the parish was considered 'High' church, but Densham was 'Low' church, and he had a definite leaning toward the Methodists, which upset the more staid members of his congregation. Within two years of his induction, the Parochial Church Council were making complaints about Densham's strange behaviour to the appropriate church authorities, with the aim of getting him removed. A petition was raised by the Parochial Church Council detailing the complaints against Densham and which document was later presented by Mr Nicholas Bunt to the Bishop of Truro.

A number of charges were laid against Densham. He was considered to be too authoritarian; he closed down the Sunday school, threatened to sell the organ and a memorial to First World War servicemen, he earmarked a piece of land on which he wished to be buried and he declined to hold services at times convenient for the congregation, which dwindled further. Eventually no one came to worship at St Bartholomew's church. Densham recorded on numerous occasions 'that there was no congregation'. To make matters worse, the Parochial Church Council declined to attend meetings called by their vicar. This situation was aggravated when, on occasion, after conducting services in his empty church, Densham walked the short distance to attend the chapel as one of the congregation. Occasionally he was invited to take part in some of the services, where he seemingly made an impression. He urged the congregation to abstain from attending theatres, cinemas or dancing, the reading

At Warleggan the Reverend Frederick Densham regularly conducted services and preached his sermons, Sunday after Sunday, in an empty church. Some visitors say that at dusk they have seen the ghostly figure of Densham moving among the trees and tombstones, his white surplice wafting in the breeze.

of novels and more of which he did not approve, saying that they were on the road to hell. In April 1949 things came to a head and Densham had a visitation from the rural dean and the Archdeacon of Bodmin, quickly followed by that of the Bishop of Truro. Apparently Densham had been quietly working away, alone and undisturbed, painting the interior of St Bartholomew's church in garish shades of yellow, blue and red. He was ordered by the bishop to engage a local tradesman, at his own expense to reinstate the interior of the church in plain white as formerly.

After an enquiry, the bishop could find no grounds for removing Densham. By this time the Parochial Church Council had had enough and as a whole they resigned. According to the *Cornish and Devon Post* of 31 January 1953, even though there was never any congregation, Densham had, over a long period, continued to hold regular services during which he sang hymns and preached his sermons in the normal manner, as if his church had a full congregation. On rare occasions, curious visitors came to Warleggan to hear Densham recite the church offices. The Reverend Frederick William Densham was a vegetarian, which in 1931 set him apart and marked him out as peculiar. By all accounts he only ate one meal a day, the main dish of which was porridge, followed by vegetables and boiled stinging nettles! Densham was undoubtedly an eccentric, and over time his odd behaviour ordered his life. The vicarage was barricaded behind a tall wire fence against enemy invaders. For added security he fixed additional locks to the vicarage windows and doors. By all accounts each door in the vicarage had a cross painted on it. The rooms were decorated in a bizarre fashion, depicting in vivid colours a particular biblical country.

Few people were allowed to set foot in the rectory other than by written appointment. Two ladies who did were Miss Mutton of St Ive near Liskeard and her aunt, who lived at Mount, just beyond Warleggan, according to a reader's letter from Miss Mutton published in the *Western Morning News* in 1973 and recalling a visit made in the early 1930s. On arrival at the gate the ladies found it locked and barred and barricaded with barbed wire. To gain the attention of the Reverend Densham, visitors had to bang a gong which looked similar to a dartboard, which drew the attention of the dogs and eventually Densham appeared. After attending a service in the church, the visitors were escorted by Densham to the rectory for tea. On the way they noticed a playground for children where children never played; there was also a cat sanctuary enclosed by wire, with a stairway for the cats to go to bed, but there were no cats. Once inside the rectory they climbed staircases and passed through doors, Densham locking each one as they progressed. Tea consisted of cake and black bananas. Miss Mutton remembers a room where many bottles of stinging nettle wine were stored. Seemingly Densham fully expected the full weight of the war to come to Warleggan and in readiness he turned the rectory into a makeshift casualty centre. There were boxes of candles, dozens of boxes containing jars of Vaseline and a large number of bandages. Legend has it that Densham also prepared to receive a number of evacuees; but none ever came! Miss Mutton said that she would not like to have the experience of meeting the vicar of Warleggan again!

This seems to be the point where fact and fiction intermingle to embellish a story. The strangeness of Densham is perpetuated in countless articles, films and books. Some writers state that Densham preached his eloquent sermons to cardboard and wooden figures in the rows of pews. According to a booklet in the church this is patently untrue and the myth has been shattered. Densham only placed black-edged memorial cards giving the names of past vicars of the parish in the pews. This sorry state of affairs persisted, however, for around twenty years up until Densham's death in 1953, which was only discovered when a handful of parishioners noticed that the usual column of smoke was not issuing from the vicarage chimney. After gaining entry they found the Reverend Densham, aged eighty-three years, lying dead on the stairs; obviously he had been making his way to bed.

Densham was later cremated at Efford in Plymouth, and in accordance with his wishes his ashes were scattered there. He left strict instructions that there should be no fuss, no mourning and no flowers. Sadly only one mourner followed his coffin, Mr Leslie Winder, a solicitors' clerk who represented his firm, John Petheybridge and Sons of Bodmin. He also represented Densham's elderly brother, Edgar Densham, who was unable to travel to Cornwall. There was only one other clergyman present, the Right Reverend William Elwell, Rural Dean of Bodmin. There were a handful of journalists from international magazines. The service was held in St Martin's church at Liskeard where the canon J. H. Parsons officiated. The *Cornish Guardian* reports that Canon Parsons had specifically requested that Densham should not be cremated in Plymouth without a funeral service first being held in Cornwall. There are now only a few people left who knew or remember the generosity of the Reverend Densham. He spent all his money on either his church or the vicarage. The truth of the circumstances which triggered the problems between Densham and the Parochial Church Council and his parishioners is gradually fading with every passing year.

Densham also painted the nearby band room in vivid colours. It is here that something of a mystery appeared. During the course of the work on the interior of the building the name John Ferris was revealed, written in large capital letters on a whitewashed wall. Who was John Ferris? That is a mystery and, at the time of writing (April 2011), despite a note pinned to a small piece of board near the church asking for information about the gentleman, the mystery is no nearer to being solved.

Another mystery engulfs Warleggan; it concerns the 'Warleggan Twinning Association', which has twinned the parish with Narnia, the mythical land created by C. S. Lewis. There is no official 'Warleggan Twinning Association' and no one knows exactly when the sign on a road junction a short distance from the church appeared. Several people have been intrigued and, despite discreet enquiries in the area and some residents being strongly suspected of being the culprits, the accusations have been strenuously denied, so the mystery remains. What is certain is that the sign is jealously guarded by Millie, a black-and-white collie dog who makes her presence heard!

On a bright, sunny Sunday afternoon in early spring with the church organ playing and the congregation singing heartily, with the rooks cawing raucously while building their nests in the surrounding treetops, Warleggan is a pleasant spot. On a gloomy day it is easy to imagine that a different atmosphere pervades! Some say that the ghost of the Reverend Densham lurks around the church and vicarage. Whether or not this is true must be for the visitor to perhaps experience and judge!

Since earliest times, almost every river valley had its mining operations and pollution was always a hazard. The valley of the River Bedalder, a tributary of the Fowey, was no exception. In January 1904 the *Cornish Guardian* reports that arsenic from Treveddoe or

Millie, a collie dog, jealously stands guards over the 'Warleggan' sign and seems to be daring the photographer to pick a handful of daffodils at the entrance to the village, which shows that, mysteriously, the village is twinned with Narnia!

Cabilla mine – a short distance from Warleggan and from which workings tin, copper and arsenic were extracted in vast quantities – had polluted the River Fowey. Water samples were taken on behalf of the river conservators and the Treveddoe Mining Company. According to papers examined at the time, water samples were taken from different locations and sent to the county analyst for examination and their views sought. Lengthy discussions took place as to the difficulty of the company and the samples of water from the different locations. Seemingly water from the pulveriser and tanks was suspect. A Mr Sachs reported that he had found no traces of arsenic in the water from the pulveriser. Proof of the adverse effects on the river and the fish in it caused by effluents from the mine workings hinged on whether or not polluted water had flowed directly into the river and in what quantities. If large quantities of polluted water were discharged into the stream then the mine would have to take the necessary steps to eliminate the pollutants before the water was allowed to reach the river. The clerk said that he was satisfied that the water was poisonous. Seemingly there was no problem when the river was in flood or when it generally ran high but only when it was low; then pollutants flowed from one pool to another and settled. It was suggested that settling tanks be built and to ask the mine officials what steps they were likely to take to obviate any future problem which may cause future pollution.

Cardinham, a Celtic name deriving from *caer*, meaning camp, and *dinan*, meaning fortress, is one of the largest parishes on the southern side of Bodmin Moor. Cardinham Castle, dating

General view of Mount village and the war memorial.

A group of local men at Mount returning home after what appears to have been a successful day fox-hunting.

Above left: St Meurbred's church at Cardinham. It is believed that St Meurbred was buried here and that his body possibly still lies in the church.

Above right: Interior of Cardinham church, where some windows were damaged by bomb blasts in 1942 and later replaced by a man named Webb. One shows a shepherd carrying a sheep, another a miner with what could be a Davy lamp, while a third depicts shire horses ploughing. A modern window remembering Reginald Runnells shows St Francis with a variety of birds, a squirrel, an otter and rabbits.

A pit stop at Fletchers Bridge, where it is Cornish pasties all round for this group of Cardinham parishioners as they take a welcome break from the serious business of beating the parish bounds. The 'pasty tax' was but a pipe-dream of a government minister. (Photograph courtesy Cornwall Centre, Redruth)

from soon after the Norman Conquest was built by one Richard Fitz Turold de Cardinan, whose family were grandees in the area and remained nearby another 200 years. At first Cardinham gives the impression of being somewhat remote, but it has an airfield and an army firing range, necessitating the occasional flying of red flags from Bellarman's Tor to warn moorland walkers that the area is out of bounds.

The parish church is dedicated to Meubred, an obscure fifth-century Cornish saint, this being the only church dedicated to him. Little is known about the saint but he obviously lived at Cardinham in the fifth century and he was murdered by beheading in Rome. There is much to see at Cardinham church, built in the latter part of the fifteenth century. The three-stage tower, built of granite moorstone, is 97 feet high and is a landmark for miles around. The body of the church consists of a nave, two aisles, original oak roofs, two Norman fonts, a collection of bench ends and an oak chest which is believed to have been made from the timber of the rood screen, which was secreted away at the Reformation. There is a fine monument to the Glynn family, who lived at Glynn House from the fourteenth century until 1820. There are some interesting brasses; one near the altar remembers Thomas Awmarle, the fifth rector, and dates from 1400, making it one of the oldest in Cornwall.

In 1942 the church was damaged by a bomb ditched by a German bomber aircraft and the east window in the chancel was wrecked in the explosion. The 'new' east window shows a shepherd carrying a sheep on his shoulder, a miner with a Davy lamp, a team of horses ploughing; all typical scenes around Cardinham. A more recent window remembers Reginald Runnalls (1898–1966) of Hole Park Mount, showing St Francis with some birds including an owl, jay, magpie and swallow, as well as squirrels, an otter and rabbits.

A short distance from the church are some humps in the ground, these being the remains of Cardinham Castle, built by Richard Fitztuold, whose family name was Cardynam, later changed to Dinham.

Further downstream, Cardinham Water comes hurtling down from its source on Bodmin Moor to enter Cardinham Woods. These woodlands of mixed coniferous and deciduous trees, over 600 acres in extent, have been managed by the Forestry Commission for over eighty years. Hidden in its acres are the relics of Wheal Glynn silver and lead mine, which started work in the mid-nineteenth century and in its heyday was one of the most successful enterprises in the district. Today Cardinham Woods is a mecca for walkers, horse riders, birdwatching enthusiasts and cyclists, who can enjoy the circular walks and cycle tracks, each suited to various capabilities. Soon the Cardinham Water escapes the confines of the woodland and swells the River Fowey.

4

Treverbyn to the Glynn Valley and Lanhydrock

Fourteenth-century Treverbyn Bridge sits comfortably in its setting where the river comes down between lush, flower-clad meadows and a tree plantation, where at the time of writing (April 2011) large swathes of trees are being felled on the steep side of the valley. Single-file Treverbyn Bridge once carried the main road from Bodmin through St Neot village to Liskeard. The bridge has two pointed arches, each of 14 feet span with cutwaters. Henderson records that at the time he was researching his book *Old Cornish Bridges and Streams*, it was in a sorry state of repair. In the intervening years it has been blocked to vehicular traffic and a new structure, slightly downstream, was built in 1929. On the riverbank immediately downstream from the newer bridge, there is a poignant granite-and-slate memorial. It remembers Eric Armitage (1923–2002) and Margaret Ann Armitage (1925–2010). Possibly the Armitages were a couple who lived nearby; they certainly loved Treverbyn and the local area. From Treverbyn the Fowey makes its winding way through an avenue of picturesque trees.

Ashford Bridge, a mile or so further downstream, is reached by a long, steep, narrow lane leading from Dobwalls to St Neot; no place to meet a circus on the move, a furniture van, a bicycle or even a pedestrian. The river greets ivy-clad Ashford Bridge, set in a tranquil glade of trees. There are stiles on either side at one end of the bridge and notices proclaim that it is private property. Leland mentioned the bridge in 1535 as having granite slabs and a 10-foot roadway. Charles Henderson dates the bridge from 1870. A more recent study undertaken by the Cornwall Archaeological Unit (now Cornwall Historic Environment Service) mentions that the bridge is shown on a map by Thomas Martyn in 1748. Ashford Bridge is not indicated on the St Cleer parish tithe map of 1843, but is shown on the St Neot parish tithe map of 1844, so the present structure probably dates from around this time. The bridge, of four equal spans with triangular cutwaters on both the upstream and downstream sides, is built of granite ashlar piers and abutments, massive irregular granite lintels spanning large dressed granite corbels carried on the piers, and is around 9 metres in length and 3 metres wide, with parapets a metre in height topped with dressed granite coping stones. Almost immediately on the downstream side the water is noticeably ruffled, possibly indicating the position of a small weir or ford which was here at one time. Further downstream is Doublebois Bridge, carrying the main Liskeard to Bodmin road, on the lower side the river tumbles over a small weir.

The Glynn Valley is notable for the number of railway viaducts, all partly hidden by dense woodland. This is the most heavily wooded stretch of its course that the River Fowey passes, as it nudges Largin, Draw, Cabilla, Laneskin, Newbridge, Hart and Great woods.

Treverbyn Bridge is, so Charles Henderson says, 'one of the prettiest bridges in Cornwall'. In 1929 the ravages of modern-day traffic became too much of a burden for fourteenth-century Treverbyn Bridge; the roadway was realigned and a new bridge built.

This photograph shows picturesque Ashford Bridge straddling the River Fowey as it flows between St Neot and St Cleer parishes.

Treverbyn to the Glynn Valley and Lanhydrock

Doublebois Bridge, which carries the Liskeard to Bodmin road over the Fowey, at the start of the Glynn Valley. There is a small weir on the downstream side.

Only a short distance away but balanced on the opposite side of the river is St Pinnock. The church is fourteenth-century. Some years ago the church, due to falling congregations, was threatened with closure. Fortunately this never happened, though today church services are limited.

By Trago Mills the Fowey flows swift and wide, chattering over a series of small waterfalls. It is hard to believe that in the middle of the nineteenth century, here by the Fowey a flourishing gunpowder factory was operated by the Herodsfoot Powder Company. To minimise any damage in the unfortunate event of an explosion, the area was heavily clad with trees to maintain a damp atmosphere. The Liskeard-based *East Cornwall Times* reported on 22 July 1865, as did the *The West Briton* the same week, that on Wednesday afternoon the Trago Powder Mill blew up. Apparently at the time there was about a ton of powder on the site. The roof of the mill was blown off and the end walls were blown down. Fortunately, two men who were working in the coopers shop about 80 yards away escaped unharmed. By all accounts the explosion was heard in Liskeard some 5 miles away and throughout the surrounding district. Today, Trago Mills is a highly successful shopping complex, where it is said that, like Harrod's, it is possible for customers to buy almost everything. One thing is certain; if you can't buy what you want at Trago Mills you probably can't buy it anywhere!

It was all started many years ago by Mr Mike Robertson, and despite numerous objections from the County Council at the time, Trago flourished, largely due to the doggedness of Mr Robertson, who was determined to succeed. At one time it was something of a 'sport' to read in the local press of the council's latest objections and Mr Roberstson's robust ripostes when he took the bureaucrats on. Today, the name of the winner is plain for all to see! One lady who lives in Liskeard makes no secret of the fact that she has, for several years, visited Trago on an almost daily basis, not always to buy but often to look around and to see what's new!

By the edge of Trago Mills, the river flows under Bodiniel Bridge and reaches Half-Way, which is one of the most popular hostelries in the Glynn Valley and is, as its name states, half-way between Liskeard and Bodmin. In May 1999 the inn was gutted by a fire, causing over £300,000 worth of damage. Since that time Half-Way has been completely rebuilt.

For much of its course through the Glynn Valley the river is closely monitored by the highway and the railway, all hemmed in by acres of dense woodland which crowd the numerous viaducts. Fire is always a hazard. In May 1949 the *Cornish Guardian* reported that there were two outbreaks in one day. Deep concern was expressed for the safety of an extensive plantation of young trees planted by the Forestry Commission. The Bodmin Fire Brigade was called in the early evening and fortunately they were able to extinguish the outbreak within an hour of receiving the call. A second outbreak occurred about two hours later; both were in thick undergrowth. It was believed that the outbreaks were caused by sparks from passing locomotives.

Just prior to Bodmin Parkway station (still fondly called Bodmin Road by the locals despite its name-change in 1985) the River Fowey flows through woodland and below Wainsford House and Wainsford Bridge, sometimes Wennaford Bridge. In May 1904 the Bodmin District Council received a letter from S. W. Jenkins Civil Engineers of Liskeard,

Single-file Bodiniel Bridge, tucked away among woodland in the Glynn Valley.

The popular 'Halfway' Inn, whose name is self-explanatory, being halfway between Liskeard and Bodmin in the Glynn Valley.

with reference to the condition of Wainsford Bridge. It seems from a report in the *Cornish Guardian* of May 1904 that on 27 October 1903, a severe flood swept down the valley and damaged the bridge. Apparently a strong eddy had formed just upstream and washed away the foundations of the central pier before falling upstream, even though when it was originally built the civil engineers had ascertained that the foundations were 4 feet below the bed of the river. Half of the pier was washed away and carried portions of the other arches with it. The stability of the remaining bridge piers, according to the engineers, cannot be ascertained until there is a fall in the water level, but they will have to be taken down and rebuilt using as much of the original stones as possible, any additional stone coming from a nearby quarry. They did not think that the foundations could be taken down much further than the original. The piers would need to be protected by an apron of stone and the arches should be paved underneath.

Discussion took place among councillors: Mr Beswetherick said local opinion was against the bridge from the start, not large enough for the volume of water, particularly in winter flood-time. Mr White suggested that the central pier be removed, which proposal was supported by Mr Laity. The cost of rebuilding would be £200, including any contingencies. As to the paying, argument took place as to whether it was a county or district bridge. During the rebuilding programme a temporary wooden bridge would be erected at an additional cost of £20. The council advised that the surveyors and the members of the various parishes carry on the traffic at once to get all immediate necessary work done.

Glynn House stands overlooking the valley of the River Fowey. By all accounts the family who finally settled here around the time of the Norman Conquest took the name Glynn

Wainsford Bridge was washed away by floods in October 1903, but was later rebuilt and is now topped by handrails. Here the bridge stands in a wooded glade.

from the old Celtic word for narrow river valley. The Glynn family of Glynn died out in the fourteenth century and the original house passed to the Carminow family and then afterwards by heiresses to the Courtnay family. Later by a strange quirk of fate, John Edward Glynn, possibly a distant member of the original Glynn family, was able to purchase the estate and in 1805 he built a new house. Seemingly another Glynn, an uncle, had designs on the house for himself. His chance appears to have come when Edmond John, a somewhat strange character, developed an odd mode of behaviour; he refused to speak to anyone and only communicated in writing. Later Edmond's uncle had him declared a lunatic and he was committed to an asylum. The uncle then snaffled the house for himself, but only lived in it for a comparatively short time. In November 1819 a fire broke out and the house was gutted, only the outer walls surviving. Several years later, Sir Hussey Vivian, recently returned from Waterloo, bought the shell and restored the house to something resembling its former glory and also added a new entrance portico and the huge granite columns to the west frontage. The Vivian family lived at Glynn until 1947 when the house and estate again went into decline for around twenty years, and its future looked bleak. In 1962 it was saved from demolition by Dr Peter Mitchell, who by all accounts was living in a nearby cottage and by chance met the estate agent. They took a look at the house, which was in a sad state of repair but, undaunted, Dr Mitchell, a biochemist from Edinburgh University, could see its potential and thought that it would make an ideal place for a research institute. Shortly afterwards the deal was signed and sealed. Over the next two years the long-empty property was restored. Not surprisingly, many imaginative stories were bandied around at the time concerning the research which was being undertaken: among the rumoured practices were germ warfare research and vivisection, probably reinforced by the fact that at night Glynn was floodlit by security lighting and tales abounded that there were armed guards around the house and guard dogs roaming the grounds! In fact the research concentrated on finding preventative medicines for a wide variety of diseases and medical conditions. Doctor Mitchell received the Nobel Prize for Chemistry in 1978.

The River Fowey has always been one of the most favoured waterways for otter hunting in the West Country. In August 1904, seemingly at the invitation of Col Edward St Aubyn, the entire pack of Dartmoor Otter Hounds were invited to come to Glynn House to stay for a sporting 'holiday' period, which would enable them to hunt extensively on and around the River Fowey.

The *Cornish Guardian* reports that there was a good turnout on the first morning, which was rather stormy. The meet was at Bodmin Road Bridge and the master was Mr A. Pitman. Reportedly there was a field of around 250 followers. The pack started off down Lanhydrock Drive and about a mile below the bridge were cast to the River Fowey, drawing nearly down to Restormel. The hounds started on a good scent around the island and were whipped back upstream to Bodmin Road station. Later they drew up to Fletchers Brook through Glynn, where there was a good drag to a field and then back through Lady Wood to Glynn Ponds and back to the station. At the county bridge the hounds gave a pretty piece of music and the river was hunted carefully all the way up to New Bridge lodge without a find. Eventually the hounds were called off and went back to Glynn kennels.

In March 1885 there was great excitement where the branch line to Bodmin leaves Bodmin Road (now Bodmin Parkway) station. After months of wrangle, delay and complicated engineering work, the River Fowey had been necessarily diverted from its original course. The new viaduct would connect and be partly embedded in lofty embankments, the approaches to these, one on either side of the river, being over 300 feet long; the six piers would be 50 feet high, their foundations set in pits 16 feet deep, the bottom fathom consisting of concrete formed of Portland cement, clean sand and stones carrying the line over the River Fowey; it would be the only viaduct of its size needed on this section of the line. The viaduct foundation stone would at last be laid and the railway line would join Bodmin to the Great Western Railway system.

The honour fell to Mrs Preston J. Wallis. Contemporary reports in both the *Royal Cornwall Gazette* and the *West Briton* state that among those present were a large gathering of local dignitaries, specially invited friends and railway officials, including Mr Bailey, stationmaster at Bodmin Road, and numerous workmen. To reach the selected pier a wooden stage had been erected, above which the foundation stone was suspended bearing the inscription, 'This stone was laid by Mrs Preston J. Wallis 5 March 1885.' Speeches were given and Mrs Wallis was then conducted to the stone and, using a silver-and-ivory trowel, spread the mortar and, with a mallet, declared the stone to be 'well and truly laid'. Evidently much cheering, from all those present, to the success of the undertaking echoed through the valley. To the amusement of those present, the workmen synchronised their cheers with the regulated whistle of an engine. The line to Bodmin was formally opened in May 1887.

Glynn House, pictured in 1906, has a colourful history. The mansion was ravaged by fire in November 1819 and has subsequently been rebuilt.

This bridge over the Fowey by the entrance to Bodmin Parkway railway station was a favourite meeting place for packs of otter hounds when they were on 'holiday' in the Glynn House kennels.

Bodmin Road station was, according to the *Cornish Guardian* of 24 November 1905, crowded with those from Bodmin, Lostwithiel and Liskeard who were of the Liberal persuasion, eager to greet Lord Rosebery on his tour of the West Country. Seemingly the down platform was decorated with flags, pampas grass and evergreen foliage from Lanhydrock. The word 'Welcome' was picked out in white lettering on a red background and prominently placed, as were numerous posters all expressing similar greeting; some bore the legend, 'Welcome to the Earl of Rosebery to our county.' Hearty cheers were raised from the crowd as Lord Rosebery appeared at the door of his special carriage. He was officially greeted by the Hon. T. C. R Agar-Robartes and Mr A. Liddicoat, who read an address on behalf of the Lostwithiel Liberal Association. Flags fluttered over the road through which His Lordship and Mr Agar-Robartes were to travel in an open landau to Lanhydrock. By all accounts His Lordship enjoyed the drive through the grounds to the house, where he was received by Lord and Lady Clifden. It is also reported that a large number of those who had gathered at the station also made their way to Lanhydrock. During his stay at Lanhydrock, Lord Rosebery planted a copper beech tree near to that planted by Right Hon. W. E. Gladstone on a previous visit to the house.

It was here, in 1974, that one of the most daring crimes anywhere along the course of the River Fowey took place at Bodmin Road (now Bodmin Parkway) railway station. The event has entered history as 'Cornwall's Great Little Train Robbery'. The drama began to unfold in the early hours of one July morning at Bodmin Road station when four men suddenly appeared on the platform. Teenager John Vanderwolfe, who had started work at 3 a.m., was the only member of staff on duty at the isolated station that particular morning. The four men were George Peter Brown, a self-employed wholesaler; Michael Richard Maloney, unemployed plasterer; John Christopher Brown, self- employed plasterer and Paul Conteh, brother of the celebrated boxing champion. All hailed from Kirby, Liverpool. By all accounts they all came with the sole purpose of robbing the mail train.

Once the nightly train from Paddington pulled away toward Penzance, John Vanderwolfe began working further along the platform, moving mailbags nearer to where they would shortly be collected by a Post Office van from Bodmin. The first John Vanderwolfe knew that something was wrong was when he was suddenly surrounded by the men; apparently they had been hiding behind some nearby railway buildings. After a short altercation and much shouting by John Vanderwolfe, a mailbag was thrown over his head and he was trussed up so that he could scarcely move. Mailbags were ripped open and their contents strewn on the platform. Other full mailbags were thrown on top of John Vanderwolfe. By this time the robbers realised that the mail bound for Bodmin that night happened to be on another train and that their previous information had been wrong and they made their hurried, empty-handed escape.

John Vanderwolfe heard the vehicle, a red Ford Capri, drive away at speed. Seemingly he had particularly noticed the parked car because it was not parked in the usual area favoured by those who regularly travelled overnight and left their cars at the station. Later John Vanderwolfe managed to free himself and alert the police by telephone. All four men were charged when it came to Bodmin Crown Court, due largely to the outstanding evidence given by John Vanderwolfe, incidentally a dedicated Special Police Officer, who gave exact details of the clothes they were wearing and, crucially, of their getaway car.

A steam train in full cry roars through the Glynn Valley in the 1950s. Bodmin Road station, now Bodmin Parkway, is in the background. (Photograph courtesy of Peter Gray from Colin Barratt collection)

At isolated Bodmin Road station, now Bodmin Parkway, an attempted mail train robbery took place in 1974. Since that time the incident has become known as 'Cornwall's Great Little Train Robbery'.

From Bodmin Parkway the river wends its way and shortly reaches the fifteenth-century Respryn Bridge with its five arches of various sizes and dates, the main pointed arch being 15 feet in span and the other round. The arches spring from outposts some 5 feet above the water. One of the earliest references to a river crossing at Respryn Bridge, set amid its tree-clad valley, comes in the twelfth century when a charter mentions that a chapel stood here. By 1300, Henderson records that fishing rights on the river belonged to the Lords of Restormel Castle. During the Civil War Respryn Bridge was a strategic crossing and the objective of some fierce fighting between the rival forces of the Royalist and Parliamentarian troops. In August 1644 Essex, who had earlier staged an outpost at the bridge, was forced to abandon it, so giving free access to King Charles, who was at Boconnoc, and Sir Richard Grenville at Lanhydrock.

In September 2006, several thousands of pounds worth of damage was done to the Resprynn Bridge when a 40-ton articulated lorry struck the bridge and a 10-metre length of the wall from road level up to the parapet was thrown into the river. The bridge was closed to all vehicular traffic for around a month while repairs were carried out.

The River Fowey flows under Respryn Bridge and caresses Lanhydrock House, which, with its long and colourful history, is in the top ten of Cornish houses. Lanhydrock originally belonged to the Priory of St Petroc in Bodmin. After the Dissolution of Monasteries Lanhydrock was bought by the Glynn family and later it passed to the Lyttleton family. In

Respryn Bridge spans a peaceful stretch of the River Fowey. Here there are two walks, one to Lanhydrock House and another beside the river. Interestingly, close to the bridge there is a designated dog bathing pool.

1620, Sir Richard Robartes bought Lanhydrock but strangely he never actually lived there. He did, however, start building the house to replace the monastic buildings but died in 1634 and it was left to his son John, who completed the scheme between 1635 and 1642. The house is early seventeenth-century and was originally in quadrangular form with embattled walls, and the gatehouse was built in 1658. In the seventeenth century the deer park stretched over some 250 acres.

At one time during the sixteenth century the Robartes family were prominent merchants and bankers based in Truro. In 1616 Roberts was knighted after paying £10,000 for his peerage and the original family name was changed to Robartes. Lord Robartes was an ardent Parliamentarian and on the outbreak of the Civil War he garrisoned the house for the Parliament. In August 1644 Lanhydrock was captured by Royalist forces under the command of Sir Richard Grenville. Later Lanhydrock became the headquarters of the Earl of Essex and also of Prince Maurice. After Lord Robartes sailed with the Earl of Essex from Fowey to Plymouth, the king granted Lanhydrock to Sir Richard Grenville, who for a short period was entitled Baron of Lostwithiel. Lord Robartes was soon to become disillusioned with Cromwell and his revolutionary idea and shortly after the Civil War he returned to the relative peace and tranquillity of Lanhydrock. In 1642 he was created Baron Robartes of Truro. After this time he took no further interest in politics. In 1679 he was created Earl of Radnor and died in Chelsea in 1685, when his body was brought home for burial at Lanhydrock.

Lanhydrock House is the most popular National Trust property in the region. In spring the carpets of bluebells are a sight to behold!

Later the mansion came to Mary Vere Robartes, sister of the 3rd Earl. Mary later married Thomas Hunt, whose second son inherited the house and estate. During Hunt's tenure major changes were made to the house. The east wing was demolished, leaving three sides of the original quadrangle. This demolition explains why at first sight to visitors it seems somewhat strange that the gatehouse stands alone. The earldom died out in 1764. Mary Vere Robartes had married Thomas Hunt and her third son, George Hunt, was to succeed her at Lanhydrock. On Hunt's death in 1798 Lanhydrock was succeeded to his niece Anna Maria Hunt, who in 1804 married Charles Bagenal Agar, third son of 1st Viscount Clifden. It was their son who in 1822 took the additional name and arms of Robartes. In 1839 he married Juliana Pole-Carew of Antony. He was Member of Parliament for East Cornwall in 1847–68 and was created Lord Robartes of Lanhydrock in 1869.

One of the most tragic headlines in the *Cornish and Devon Post* of 1881 appeared on 9 April that year. It read, Lanhydrock House, residence of Lord Robartes and one of the most beautiful ancestral houses in Cornwall, has been burnt down. It seems that the fire quickly caught hold and reached alarming proportions in a very short time. The outbreak was discovered in the early afternoon by several people, including Lord Robartes himself, who happened to be in the grounds when he noticed smoke billowing from the house. Apparently a number of hot water pipes which run through the house had recently been made more efficient. Two new ones had recently been installed and had only been working for about two weeks. It is believed that, as the fire in the kitchen was constant and the fire was first spotted in the room immediately above the kitchen complex, somewhere in the flue a beam had been exposed and had probably been smouldering for some days. Conjecture has it that the fire penetrated the chimney and later the roof, from where it rapidly ran along the roof timbers. Seemingly the rapidity of the blaze was caused by the fact that under the roofing slate there was tarred felt and woodwork above the ceilings. Immediately the outbreak was discovered messages were sent to Bodmin and Lostwithiel. About twenty soldiers with their fire engine came from the Bodmin Barracks. Additional appliances also came from Bodmin and Lostwithiel.

Although the appliances came in record time the fire had already gained a stronghold on the roof. The rescue operation, led by the Bodmin Brigade, was under the direction of Mr W. H. Buscombe, assisted by Mr Henry D. Foster and Mr Burnard Edyvean. It was hampered by the shortage of water, the only supply available being from a tank used to supply an ornamental fountain. The fire was burning its way toward the north wing of the house, which contained the long picture gallery. Fortunately it was saved and the 116-foot-long and 18-foot-wide gallery on the upper floor was saved. On one side there are four windows and three on the other, one being blocked up on the inside when new library shelves were installed. The barrel ceiling, plaster star-shaped panels in high relief, show episodes from the Old Testament. Smaller panels depict fabulous animals and real birds. The Great Hall has numerous family portraits by the eminent artists of the day, including Lely, Kneller, Romney and Hudson.

At one period, the dry shrubs in the narrow space between the main house and the church caught alight and threatened to overwhelm the ancient church. Quick action by some of those present saw the shrubs cut down and water was trained on the roof of the north wing, and late in the afternoon the fire was prevented from reaching this part of the house and the church. Unfortunately the other side of the quadrangle, due to the strong wind, burnt fiercely. An operation to salvage several pieces of valuable furniture, a number of which had to be taken roughly apart, swung into action, the upright bars of the Elizabethan windows being an impediment. Many pieces were removed by this method. Many priceless paintings and countless books from Lord Robarte's library were carried from the house to the safety of the church, the fire's spread having been successfully halted.

The fierceness of the fire is shown by the fact that two maidservants who were working upstairs at the time of the outbreak were trapped and were rescued through the windows using ladders. By all accounts the fire ripped through the servants' apartments and they lost all their personal possessions. Early in the evening the roof had collapsed and the whole of the south wing and the western end of the house was open to the sky. Some ceilings and woodwork remained but the gusting wind threatened to fan the flames. Later the fire's spread was halted and confined to the south wing; in other parts of the main house the fire was left to burn itself out. The newspaper reports graphically tell how in the darkness the scene was one of magnificent grandeur, the wind fanning the flames, which were licking through some of the pane-less windows, strangely lighting the mansion and its surroundings.

Soon after the disaster Lord Robartes appointed Cornishman Richard Coad from Liskeard as his architect for the building. Only stone, granite and roofing slates from local quarries were used. Greatly shocked by the fire, Lady Juliana Robartes sadly died within a few days, Lord Robartes passed away the following year. Undaunted, Charles Robartes, who became the new Viscount Clifden, set about rebuilding and enlarging the house, which is much as we see it today. Obviously fire prevention was uppermost in Lord Robartes' mind at this time. Wrought-iron girders were used instead of timber, concrete floors were laid and the ceilings were made from an especially fireproof composition. Mains water was brought through cast-iron pipes to a private reservoir and hydrants conducted the water supply to every part of the house, both interior and exterior. A hot-water boiler and a central heating system were installed, as well as electricity. The work was eventually completed in 1885. The Robartes family made Lanhydrock their seat from that time until 1953 when the seventh Viscount Clifden generously endowed the house and estate, comprising well over 400 acres, and gave it to the National Trust. He continued to live at Lanhydrock with his sisters until his death. His last surviving sister, Lady Everilda Robartes, died in 1966, bringing to an end the long family occupation of the house. There are a range of stables, kitchen gardens and formal gardens. In season the ocean of bluebells are breathtaking, as are the rhododendrons and magnolias.

Over the decades Lanhydrock has played host to several royal visitors and politicians. Lord Roseberry and Mr Gladstone both planted copper beech trees. The majority of the ancient sycamore trees planted in the avenue in 1648 to mark Cromwell's victory still survive, although replacements for some have been augmented with beech trees. Leaving Lanhydrock behind, the growing River Fowey flows on toward Boconnoc and Lostwithiel.

5

Braddock,
Boconnoc House to Lostwithiel

Braddock, Bradock and Broadoke are only some of the many and varied spellings of this parish, squeezed between St Pinnock, Boconnoc and St Winnow and separated from Cardinham, Warleggan and St Neot by the River Fowey. The church, standing somewhat forlorn, is dedicated to St Mary. The two-stage tower has stump pinnacles and the body of the church comprises chancel, nave, north transept and south aisle. The wagon-roof timbers remain. At one time the top part of the rood screen was removed to Boconnoc while the lower section remains. During the Civil War, in 1646, here in the vicinity of Braddock church the opposing Royalist and Parliamentary forces faced one another before the fierce Battle of Braddock Down. Although standing in a vulnerable position, Braddock church stood firm. Seemingly on occasion there is a sense of not being entirely alone. By all accounts soldiers wearing bedraggled uniforms of the Civil War period, with cannon, muskets, pikes and shields, have been seen and the clash of steel on steel heard! One lady who visited Braddock church many years ago swears that when she entered the churchyard, she saw Cromwell on horseback. Oddly, her companion did not!

Boconnoc House (pronounced BO-CON-NOCK) and its estate, with its deer park, cricket pitch and a sizeable lake, together with a number of farms and cottages, spreads itself amply over several thousand acres, mainly on the east bank of the River Fowey a few miles from Lostwithiel. Boconnoc, which dates from the time of the Domesday Book, has a long, proud and colourful history. Over the centuries Boconnoc has belonged in turn to a chain of eminent families, including the De Cantcia, Carminow, Dawney, Courtenay, Russell, Mohun and Pitt families. In 1864 it was bequeathed to George Matthew Fortescue and his descendants still live at Boconnoc. Looking at Boconnoc, one thing is certain; the house and its occupants are inseparable from both National and Cornish history.

The Civil War came to Boconnoc when the Battle of Braddock Down was fought a short distance away. In 1644 Royalist forces encamped here and the royal standard was stuck in the trunk of an old oak tree. On one occasion, so the story goes, when King Charles was taking the sacraments under the branches of the oak, an assassination attempt was made on his life. Seemingly a musket ball narrowly missed the king, passing through the tree trunk. Tradition has it that from that time, the tree has been stunted and has borne mottled leaves. In the late nineteenth century, according to Miss M. A Courtney in her book *Cornish Feasts and Folklore*, a hole in the tree trunk was pointed out to visitors to confirm the story!

A priceless diamond is woven large into the fabric of the Boconnoc story. In 1701, during the time when Thomas Pitt was Governor of Madras, he purchased a large diamond, valued at around £20,000, from a reputable Indian diamond merchant. This stone, due to its size

Above left: St Pinnock church was dismissed by Nikolaus Pevsner as being 'a church of little importance'. Inside is a Norman font with corner heads and arms supporting a heavy, wide square top, old timbers and a wagon roof. St Pinnock was once destined for closure due to lack of congregation. Services are still held at St Pinnock, so evidently congregations improved and prayer prevailed.

Above right: Surrounded by trees, isolated Braddock church miraculously escaped severe damage even though the Civil War raged in the area. Inside there are interesting features, including the pulpit, wagon roof and bench-ends. It is believed that one of the pinnacles came from St Nectan's chapel at St Winnow.

and weight, became known as 'The Great Pitt Diamond' and afforded Pitt the nickname Diamond Pitt. In 1717, Pitt sold the diamond to Phillipe d'Orleans, Regent of France for £134,000. From this windfall Thomas Pitt bought Boconnoc House and estate for £54,000. Interestingly the diamond was later set in a crown worn by Louis XV at his Coronation and sometime later set into the hilt of a sword belonging to Napoleon. Pitt settled at Boconnoc and in 1717 added an additional wing to the house. During his time at Boconnoc Pitt had two sons, Robert and Thomas. Later two grandsons followed; William became the 1st Earl of Chatham and Thomas was elevated to Baron Camelford in 1784 when his cousin William Pitt became Prime Minister. In 1720 Thomas Pitt extended the house, and further additions were made by his grandson. In 1771 another Thomas Pitt created the long gallery and made further extensions to Boconnoc House. It was around this time that he designed and laid out the parkland. Seemingly Boconnoc is always evolving. In 1834 the north-east wing was demolished, an Italian-style window was added to match an earlier window fitted into the east front, and in 1865 the tower was added.

Boconnoc House, the home of Thomas Pitt, known as Diamond Pitt, is dripping in history from every century. This image shows the crowds at the Cornwall Garden Society Flower Show in 2011. It is said that Boconnoc has over 20 miles of private drive within the park, which is the largest in Cornwall.

Not surprisingly, Boconnoc was involved in mining in the Lostwithiel area. A supposedly rich vein of silver lead was struck but apparently its actual output fell sadly short of its forecast potential. In March 1868 the *West Briton* reported that in the previous eleven years over 600 acres of wasteland had been reclaimed at the Hon. G. M. Fortescue's estate at Boconnoc.

During the Second World War most of the house and all the outside buildings were taken over by American troops who were stationed at Boconnoc. There were large ammunition dumps in the park in preparation for 'D' Day, Fowey, an important embarkation point, not being too far away! The house sustained damage during this period and suffered from general damp and decay. In 1969 an architect's report raised serious concerns about the house, and in 1972 the weighty decision was made to demolish the south-west wing.

Boconnoc easily lends itself to the sets for period films. Several, including *My Cousin Rachel*, from the Daphne Du Maurier novel, and adaptations of the books of Rosamund Pilcher by a German film company, have been filmed on the estate. The world of television has not overlooked Boconnoc either: numerous scenes from the evergreen television series *Poldark* from the Winston Graham series of novels set in eighteenth-century Cornwall have made use of the location. Boconnoc House became the home of George and Elizabeth Warleggan in the second television series as fictional Trenwith was burned down at the end of the first television series.

In 2001, after being neglected for around thirty years, a major restoration programme for Boconnoc was launched, beginning with the Victorian tower, followed by the roof and main house. At the time of writing (August 2012) it has just been announced that Boconnoc House has been awarded a national architectural award. The recognition comes from the Historical Houses Association, which has given Boconnoc their prestigious Sotheby's Restoration Award, after the Fortescue family undertook an eleven-year restoration project which has transformed Boconnoc.

Recently Boconnoc has inaugurated the 'Boconnoc Music Award' in conjunction with the Royal College of Music. In this first year, the award will enable a quartet of students from the Royal College of Music to stay at the house for a week. While at Boconnoc the students will prepare and rehearse a programme, and toward the end of the week they will perform recitals and a concert in Boconnoc church.

Today, one of the most prestigious events on the Boconnoc calendar must be the Cornwall Garden Society Spring Flower Show, which at the time of writing has just celebrated its 100th anniversary. Established in 1832 as the Royal Horticultural Society of Cornwall under the generous patronage of King William IV, the society held its first Daffodil and Spring Flower Show in March 1897. The show has been held annually, with only a few short breaks since that time. Today it is estimated that over 10,000 visitors converge on Boconnoc for the two-day show when a great variety of rhododendrons, magnolias, camellias, daffodils, trees, shrubs, plants and floral displays attract competitors from all over the country and abroad.

Always moving with the times, Boconnoc is a popular wedding venue, and during the year traction engine and vintage vehicle rallies take place, along with the ever-popular Fun Dog Show! In May 2013, Adventure Running World Championship will be held at Boconnoc for the first time.

Fifteenth-century Boconnoc church peers over the hedge toward the mansion. Inside the church there are north and south aisles, a chancel and a nave. The arches are supported by huge granite pillars. Also inside are the original roof timbers, minstrel gallery and font. The altar table bears the inscription 'made by me, Sir Raynold Mohun. 1621.' There are some particularly interesting memorials.

Boconnoc church, built mainly in the fifteenth century, consists of chancel, nave, north and south aisles and minstrel gallery. The church seems to peer shyly toward the house. Its octagonal Victorian bell turret holds one bell. According to the Boconnoc guidebook, at one time the bell was housed in a small shed, the bell being rung by means of a handle similar to a water-pump, and scarcely sounded within the mansion. The turret was built at a cost of around £70 and the bell replaced. There are some fascinating monuments, including the Royal Arms of 1639, and there are windows in memory of members of the Fortescue family and a richly carved Gothic screen. The soaring 123-foot granite obelisk in memory of Sir Richard Lyttelton, 12 feet square at the base and standing aloof in the centre of an area where troops probably gathered prior to the Civil War Battle of Braddock Down, dominates one part of the park. It was erected by Thomas Pitt, 1st Lord Camelford, to his wife's uncle and benefactor in 1771. The inscription reads,

In
Gratitude and Affection
To the memory of
Sir Richard Lyttelton
and to perpetuate
that peculiar character
of Benevolence
which rendered him the
delight of his own age
and worthy the veneration
of posterity.
MDCCLXXI

The River Fowey seems reluctant to leave Boconnoc, but eventually decides to weave its way through the lush valley toward Restormel Castle. Further downstream, this venerable edifice stands proudly on a bluff of high ground, which slopes away on three sides on the west bank of the river. It is one of the gems of the river valley and a short distance before one of the main crossing points on the River Fowey. It is believed that the site was originally an earthwork with wooden palisades surrounded by a ditch.

Originally Restormel was part of the Manor of Bodardle, which at the time of the Domesday Book was held by Turstin, the Sheriff of Cornwall. It is thought that his son, Baldwin FitzTurstin, built the castle in around 1100. The Norman castle was built from stone brought from a nearby quarry. The keep would have been approached from the bailey. The massive, formidable three- storey gatehouse would have had a portcullis in front of the main outer gate and a drawbridge over the moat, all designed to make a show of strength and show who exactly was boss and had the sway hereabouts. The castle had a vast range of rooms, kitchens, sleeping quarters, a solar and a private chapel for the lord and his family. At the time the castle, built more for status, show rather than defence, would have been white. There were quarters for guards, masses of staff, stabling for numerous horses.

In 1166 Robert Fitz William Lord of Cardinham held Bodardle. Later Restormel Castle was inherited by Andrew de Cardinham and later in 1217 it passed to Isolda de Cardinham,

The obelisk which remembers Sir Richard Lyttelton is a soaring feature of Boconnoc Park. It has long been believed that the obelisk stands on the site where the Royalist forces camped before the Civil War Battle of Braddock Down. New research has thrown doubt on this supposition.

Circular Restormel Castle at Lostwithiel is the iconic castle in the River Fowey Valley. This picture shows it as it was over a century ago.

who in 1264 married Thomas de Tracy. Later in 1270 Isola De Cardinham granted the castle and land to Richard, younger brother of Henry III, who became Earl of Cornwall. In 1272 Richard was succeeded by his son, Edmund, whom history believes built the shell wall and the internal buildings and put much of the work at Restormel in train. He made Restormel the main centre of the tin industry. Huge revenues were ploughed back into improving the castle. Timber structures were replaced with stone and a vast array of service buildings were erected in the bailey. Improvements were also made in the keep; rooms were added and there was also a mains water supply. On the death of Edmund in 1299 the Earldom of Cornwall reverted to the Crown and later the Duchy of Cornwall. In 1337 King Edward III created the first Duke of Cornwall when Edward the Black Prince was seven years old. He came to Restormel in 1354 and again in 1365 and by all accounts ensured that during this period the castle was well-maintained. This line of Dukes of Cornwall continues to this day and will hopefully continue into the future.

Charles Henderson, the eminent historian, tells that when the Black Prince, 1st Duke of Cornwall, came to the county, he always stayed at Restormel. At that time Restormel was the only castle which offered what could be considered as 'five-star' accommodation, this being expected by the duke, including mains water brought by lead pipes from a spring further up the hill, and latrines. When the lord was at home or high-flying personages were being entertained the central courtyard became an arena for displays of archery, one-to-

View of Restormel Castle as it is today. Its walls are several feet thick and enclose a circular area around 100 feet in diameter. Staircases allow visitors to walk around the ramparts and admire the view.

one combat and horsemanship. The building would be hung about with standards and banners, all designed to enhance the status of the lord and also to impress visitors. In 1584, Elizabethan historian John Norden records that he was impressed by the castle layout and its obvious strength, but Restormel's attractions were fading and by Tudor times it was becoming ruinous.

During the Civil War, the castle was garrisoned by the Earl of Essex and the Parliamentarian forces while they held Fowey and Lostwithiel. At the time the large eastern window of the chapel was blocked up so that a gun emplacement could be mounted there. In August 1644, Restormel Castle was captured after a short siege by the Royalist forces under the command of Sir Richard Grenville. After this time Restormel was never lived in again and its importance waned and it gradually fell into decay. Legend has it that in the late eighteenth century some workmen were working in the cellars under the courtyard and found some armoured skeletons. Other remains have also been found. By the eighteenth century the castle had become an ivy-covered ruin and as such it fired the sometimes romantic imagination of

Looking down from the ramparts into the interior of the castle.

artists. In November 1841 the *West Briton* reported that when the news of the royal birth reached Lostwithiel, a brass cannon was taken up to the ramparts of Restormel Castle, from where a royal salute was fired. A local young lady made a royal standard for the special occasion as well as a Union Jack and the Star of Brunswick. After the salute, three hearty cheers were given in honour of the event.

Restormel Castle once stood in an extensive deer-park where it is recorded that at one time over 300 deer ranged. Though still owned by the Duchy of Cornwall, their guardianship was relinquished in 1925 when it passed to the then Commissioners of Works until 1984, when Restormel Castle came into the care of English Heritage, who today manage the site as a tourist attraction standing as a dignified ruin above the River Fowey. At a Cornish Gorsedd Council meeting in February 1971, so the *Cornish Guardian* reports, a proposal was made that Restormel Castle should be restored. Lengthy discussions took place and, reading between the lines, heated exchanges took place. Mr Dennis Trevanion said that he felt that the College of Bards was generally against 'mucking about with Restormel Castle'. Mr E. Morton Nance said 'that while the matter did not come within the Gorsedd purview he did not see anything wrong with the restoration proposal, if the restoration proposal would attract more visitors, so much the better'. Mrs D. Keighwin Ledbury said 'she would hate to see Restormel Castle spoiled'. Mr Henry Trefusis said 'the idea was ridiculous and should be allowed to die as quickly as possible'. The discussion produced no motion and the Gorsedd agreed to take no action!

Georgian Restormel Manor, standing in the shadow of Restormel Castle, is believed to have been built on the site of an ancient chapel. The house has been the home of several well-to-do families over the generations. Today it is owned by the Duchy of Cornwall and some of the outbuildings have been converted to superior holiday accommodation.

Restormel Manor House, a Grade II listed, sixteenth-century, two-storey property, seems almost in the back garden of Restormel Castle. The house is built on or near the site of an earlier chapel dedicated to the Trinity which was destroyed in the Reformation. Today the house, owned by the Duchy of Cornwall, contains luxury apartments and several of the range of outbuildings have been converted into holiday accommodation. The river glides a mile or so downstream and on toward Restormel Bridge, built in 1939, carrying the Lostwithiel bypass traffic away from the town. It was reported in the *Cornish Guardian* of April 1939 that the motor traffic through the town during that Easter had been very heavy and on the Monday it was almost continuous.

The iconic old bridge, the lowest on the river, was built by the Normans who founded Lostwithiel. The Great Bridge at Lostwithiel was mentioned in documents of 1280 when it had nine arches; the four most westerly are now under North Street. Charles Henderson tells that in June 1358 the Black Prince, Duke of Cornwall, issued an order to one John De Kenall, the keeper of Restormel Park, to take a number of men to repair Restormel Bridge, which was in a ruinous condition. Over the centuries the bridge has undergone several changes. It was largely rebuilt in the fifteenth century and in 1644 it was nearly blown up by the Parliamentary troops. In 1676 the parapets were added and in the eighteenth century four easterly arches were added after the river changed its course.

The nightmare of heavy vehicular traffic passing through the historic heart of Lostwithiel was somewhat relieved in 1939 when Restormel Bridge was opened.

In Lostwithiel history seems to crop up in the most unexpected places. The *Cornish Guardian* of October 1904 reports that workmen who were excavating in Bridge Street for a new drain opened up two arches of the old bridge which had previously been covered. Seemingly the arches were of a very ancient design, believed to date them from the time of King John. A surprise was the discovery of an old English silver dessert spoon with a shell-shaped bowl, looking as if it had been in a fire, but clearly bearing a lion's head, the letter 'F' and the initials T. N. The report states that the spoon is in the possession of Mr Oliver in Bridge Street, and he will be pleased to show it to anyone who is interested. The river emerges from under the bridge and greets Lostwithiel and the tide.

The old bridge over the Fowey at Lostwithiel dates from the twelfth century. During the Civil War the bridge was considered a strategic crossing and the Roundhead forces formulated plans to blow it up and reduce it to rubble. Today it is the iconic bridge on the River Fowey.

6
Lostwithiel and Further Downstream

After cruising down the valley toward Lostwithiel, the River Fowey emerges quietly from under the medieval bridge to greet the town. It is fair to suggest that Lostwithiel has its own special charm, a quiet friendliness bordering on humility, but there is no reason for the town to be so unassuming. Lostwithiel oozes more history to the square yard than many other Cornish towns. Once it was one of the most important places in Cornwall and, certainly before the river silted up, a port to rival Fowey. Plain to see on almost every corner of the quaint streets are the signs of this unique architectural heritage, standing cheek-by-jowl with more modern, tasteful developments. Over the centuries historians have forwarded various ideas about the correct spelling of the name of the town. In 1189 it was Lostwetell, then later a range of spellings and meanings, but today it is generally accepted as being Lostwithiel, from the old Cornish Lostgwdeyel, meaning 'place at the end of the wood'.

One of the undisputed gems of Lostwithiel is St Bartholomew's church. The main body of the church dates from the twelfth-century Early English style, on the site of a church dating from AD 542. The tower was added in the thirteenth century and during the next century the octagonal spire with its ornate lantern base, unique in Cornwall, was added to bring the overall height up to 110 feet. There is an interesting incident connected with the church spire. In October 1863, one of the more sensational events ever associated with St Bartholomew's church took place while restoration work was being carried out on the church spire. Seemingly someone had remembered that about 100 years earlier, a man named Bramble had ascended to the cross on top of the church spire and drank a pint of beer. According to the *Royal Cornwall Gazette* the present workman, a man named Saunders, was challenged to do the same! Word must have spread that the challenge had been issued and that Saunders had accepted and was eager to prove his challenger wrong, which caused quite a sensation in the town. The report continues to tell that at the appointed time a large crowd of people gathered to watch the feat. Saunders duly ascended to the cross, drank his pint of beer and hurled the pot to the ground! Minutes later Saunders descended safely to the ground amid great cheers from the assembled crowd.

For around 500 years the tower had a right of way through it but this came to an end due to developments in nineteenth century when the footpath, after some 700 years, was redirected. In the sixteenth century a great change came about with King Henry VIII's Act of Supremacy in 1534 when everyone, instead of paying allegiance to the Holy Father in Rome, had to recognise the king as the Supreme Head of the church in England. A this time St Bartholomew's, instead of being a daughter church of Lanlivery, became a parish church in its own right.

Above left: Standing back from the main road, almost every street in Lostwithiel seemingly has a building of special interest, as well as quaint corners and half-hidden courtyards. There is a fine walk by the river. This picture, dating from 1971, captures the main road at a time when it was devoid of traffic.

Above right: St Bartholomew's church, Lostwithiel, with its pencil-slender spire, has a colourful history. An interesting memorial in the porch remembers Joseph Burnett, a town sergeant who was killed on duty in August 1814. There are interesting tombstones and an old cross in the churchyard.

Inside the church there are several interesting features. The richly carved octagonal font of Pentewan stone dates from the thirteenth century and shows in fine relief the crucifixion, lions and huntsmen on horseback with the hunting horn and hawks. Close by is a poor box dating from 1645 and given to the church by William Taprell during his mayoralty.

During the Civil War Lostwithiel was staunchly Royalist. But in 1644 the town was controlled by Cromwell's Parliamentary forces and consequently several buildings, including the church, suffered a great deal of damage. By all accounts the Parliamentary forces stabled their horses in the church and on one occasion a horse was baptised in the font and named Charles after the king. At one period, when Royalist soldiers were imprisoned in the church, two climbed into the tower and refused to come down when challenged by the Parliamentary forces under the Earl of Essex. In an effort to dislodge them they tried to smoke them out without success and they resorted to more drastic means. They ignited barrels of gunpowder and the resulting explosion blew off the church roof. In January 1757, as recorded by John Smeaton at the time in the *Philosophical Transactions of the Royal Society*, the church spire

General view of interior of St Bartholomew's church. In medieval times the figure-eight was a symbol of the resurrection and the figure appears several times in the church. There are eight octagonal pillars, the lantern of the elegant spire is octagonal, the pulpit and the font are octagonal and there are eight clerestory windows.

was struck by lightning and shattered. Apparently the storm was so sudden and fierce that it frightened the town's inhabitants. Around 20 feet of the tower was thrown down and fell in all directions and many stones fell through the church roof, seriously damaging the pews.

Arguably one of the most interesting buildings in Lostwithiel, known since 1337 as the Duchy Palace but originally the Great Hall or Stannary Hall, though noted for its great size, is not a palace at all. it was originally built in the twelfth century of mainly locally quarried stone by direction of Earl Edmund on a seemingly requisitioned corner site running parallel with the River Fowey and its confluence of the River Cober, now hidden under paving stones. The Great Hall served as an administrative centre, particularly for the collection of rent on behalf of the Duchy of Cornwall. Over time the building housed the County Court and was the centre for the weighing, assaying and the storing of tin. The tinners conducted their business in the Convocation Hall and at one time the Stannary Prison occupied part of the building. During the Civil War the building was occupied by the Parliamentary forces who, under the Earl of Essex, ransacked the building, burning many valuable documents including the Stannary Parliament records as well as doing untold damage to the building. Later the Roundheads were surrounded by the Royalist forces, who forced them to escape downriver before evacuating by sea. Eventually the palace building became ruinous and in the seventeenth century it was partly demolished.

The palace served as the duchy office until 1874. Strangely, it is recorded that at one time

Surprisingly, at St Bartholomew's the fourteenth-century tracery and mullions survived the ravages of the Civil War. The stained glass east window is the largest in any church in Cornwall. Measuring 34 feet by 14 feet, around ninety figures featured tell the story of the crucifixion and resurrection in nineteenth-century glass.

At the time of writing, the Duchy Palace in Lostwithiel is shrouded in polythene sheeting as the building is undergoing an extensive restoration programme.

in the nineteenth century, part of the building served as a slaughterhouse as Lostwithiel and was an important centre from which to send meat to the London markets. Later, in 1874 the Convocation Hall came onto the property market and it was purchased by the Freemasons. At the time of writing (March 2012) the Duchy Palace has been mostly closed to the public for around 125 years and is undergoing an extensive £600,000 refurbishment programme, the first since 1878, under the auspices of the Prince's Regeneration Trust and the Cornwall Buildings Preservation Trust. The work is being undertaken by Carrek, a building firm based in Bath who specialise in the conservation and regeneration of historic buildings. On completion of the regeneration work, the Duchy Palace will be used for office accommodation and small retail units together with an exhibition room for photographs and artefacts showing something of the history of the building over 700 years.

Lostwithiel Museum was opened in 1971 by the Earl of Mount Edgcumbe, accompanied by the countess. The museum is the brainchild of members of Lostwithiel Old Cornwall Society, who after protracted negotiations with the town council were given the go ahead! Today it occupies the ground floor of the Guildhall building, which dates from 1740. At one time it was the Corn Exchange and town lock-up, which had three cells and a barred window, still in place in a corner position in one of the museum display rooms. The displays cover a wide variety of artefacts, some of which are unique, relating to the town and district. Lostwithiel's first eighteenth-century fire engine is one of the main exhibits, another is the cloak being worn by Joseph Burnett, the town sergeant murdered in 1814 during the course of his duties. A curiosity is a slice of a tulip tree, a relative of the magnolia. This tree, standing 29 metres tall, the second-tallest in the country, once stood in a private garden in the town and was cut down in October 1999 having become dangerous.

In October 1903 the area around Lostwithiel was hit by a terrific storm which resulted in severe flooding. It was reported in the local newspapers, the *Cornish Guardian* and the *Launceston Weekly News*, as 'The Great Flood, Lostwithiel'. One newspaper states, 'Not within the memory of anyone living can such a disaster be remembered. In Lostwithiel great damage done to shops, Mr Skilton grocer, Messrs Pearce, printers, Roskilly, ironmonger, Gill, outfitter, the Post Office and the Boconnoc estate office in Queen street G. R. Brown draper, J. Daniel shoe maker in Fore street, all severely affected by the torrent which hurtled through the town.'

Leaving Lostwithiel the river slides by the Brunel Quay apartments, converted from the old railway carriage works in 2011, and under the railway bridge and by Coulson Park, created in 1907 by Nicholas Coulson, whose name it carries today. There are lime trees and an array of wild flowers edging the river as it flows seaward.

Of all the Fowey creeks, Lerryn is arguably the finest. The Lerryn River comes down from beyond Couch's Mill and Boconnoc to reach Lerryn village where the single-file bridge, first mentioned in documents by Leland in 1535, marks the tidal limit of the River Fowey. A manuscript in the Bodlian Library records that in 1573 Queen Elizabeth made a directive to the bailiff and constables of the Hundred of West to levy a rate for the erection and repair of the decayed bridge called Laryon between St Vepe (sic) and St Winnow. The present bridge, so Charles Henderson believes, is possibly as old or possibly older. At first it had three pointed arches but today has only two, one larger than the other and both springing from imposts. At low tide, below the bridge a line of stepping stones over the river is revealed. The *Cornish Guardian* of February 1939 reported that at a meeting of the Council for

At Lostwithiel, these old Great Western Railway maintenance depot buildings, designed by Isambard Kingdom Brunel and standing beside the River Fowey, have been tastefully converted into luxury riverside apartments and appropriately named Brunel Quay.

The quay at Lostwithiel was always working to capacity, but some old limekilns remain and have been converted to other uses.

Lostwithiel station complex is separated from the town by the River Fowey. The main railway line came to Lostwithiel in 1859 and the line to Fowey opened in 1869. Some of the original wooden buildings remained until 1981. The station has always been busy, particularly with the milk trains destined for the milk depot, which opened in 1932. Frequent china clay trains pass through Lostwithiel en route to the docks at Fowey. (Photograph Colin Barratt collection)

the Protection of Rural England, Mrs Pollard announced that as a result of the council representations the executive committee reported that Lerryn Bridge had been scheduled as an ancient monument by HM Office of Works.

Lerryn village is picturesque and is a favourite with fishermen and artists, featuring quaint old cottages, the old forge and mill, limekilns, the eighteenth-century Ship Inn, a village shop and post office. Views down the creek, greatly expanded below the bridge, are enhanced by waterside buildings, graceful swans and boats bobbing on the water or lying idly on the mudflats. Lerryn has always been famed for its regatta and it is a great 'date' in the village calendar. The *Cornish Guardian* of August 1960 reported that competitors came from Falmouth, Mylor, Oxford and Plymouth to take part and that over 5,000 people converged on to the village. It seems that some things never change: there were grumbles among the 1,500 people over the car-parking charges – 2s (10p) for parking in Tivoli Park, from where they had a grandstand view of the river. Apparently both sides of the river were gay with fairy lights, flags and bunting. There were sideshows, highly painted funfair caravans and canvas booths. A tidal wave of nearly 18 feet set a tsunami wave rushing up the river, lapping over its banks and footpaths. Motorists who had parked on the roadway were trapped by the water.

Beside the creek is Tivoli Quay and Ethy (pronounced EETHY) House, a Georgian property adorned by pilasters, partly hidden from view by a rich, wooded parkland landscape of some 375 acres. The house has passed through several prominent families, including Stonard,

View of Lerryn across the creek from Tivoli Quay.

The Trigg Morris Men regularly dance at Lerryn, much to the delight of the villagers and visitors. (Photography courtesy Roger Hancock and the Trigg Morris Men team)

Boats are always part of the passing scene at Lerryn, which is beloved by artists and photographers. A lone oarsman is taking advantage of the tide. Note the gardens on the sloping hillside.

Kayell, Courtney, Earl of Mount Edgcumbe, and Howell. In 1798 it was leased to Admiral Sir Charles Vinicombe Penrose. Later it was occupied by the Melvill family. The property was given to the National Trust by Lord St Levan as an endowment for St Michael's Mount. The house is not open to the public though the gardens are open on special occasions.

In July 1949 it was proposed to establish a Farm Institute in Cornwall, and Ethy Barton at Lostwithiel was deemed to be the most suitable location. Initially the Ministry of Agriculture, the Cornwall County Council and the Farmers Union were all enthusiastic about the project. The purchase price of the property was £20,000 but with the necessary adaptations to the property to accommodate forty students and later between seventy and eighty, the building of roads, paths, fences, providing glass houses, livestock, implementing furniture and equipment, the cost soared to nearly £100,000. Suddenly the discussions which had been going on over several years became intense, with various viewpoints being put forward. Chairman of the Cornwall County Council Lt-Col E. H. W. Bolitho and several other councillors were in favour of the project. Financial concerns were much to the fore. Some councillors thought that the financial burden was too heavy to put on the public; others said that there was no need for a technical education for growing potatoes or broccoli, though there may be for growing strawberries! Another point of view was that the county was already doing a lot for farming and if farmers wished to give their sons and daughters a technical education then they should pay for it! Although the ministry of agriculture agreed with the establishing of a Farm Institute in Cornwall, the Cornwall County Council thought that it was not the right priority at the time and declined to proceed with the undertaking.

Lerryn Bridge marks the highest navigable point on the Lerryn Creek. Old limekilns stand nearby.

At various times, Ethy House above the River Fowey has been the home of the Stonard, Kyell, Courteney, Vinicombe and Penrose families.

St Nectan's chapel or St Nighton at St Winnow, where only occasional services are held, stands proud in its field. Research reveals that it has a fascinating history. Dedicated to St Nectan, eldest son of King Brychan, he arrived in Cornwall from Wales during the fifth century. He later became an important figure in the kingdom of Dumnonia, now Devon and Cornwall. He was beheaded for his beliefs in the seventh century. No one knows exactly when the chapel was built but what is certain is that an early mention of the chapel comes in documents dated 1250 as daughter church of St Winnow, when it had a glebe attached to it. The *Cornish Guardian* of September 1905 records that in a survey carried out in AD 1281, St Nectan's possessed a silver chalice, vestments, two blest towels, one unblest, a font and a good bell. Presumably it had all been recovered as it is stated that they were stolen by three people in autumn of 1279! In the early fourteenth century a monthly mass was said at St Nectan's as people were, by all accounts, unwilling to attend services at St Winnow. The building was remodelled in the fifteenth century when there were some 15 acres of glebe land attached. In 1613 a report recorded that the fabric of the building had fallen into decay. The vicar, too, seemingly was neglectful; apparently no services were held on a Sunday or on any other holy day for over six months, and worst of all no celebration of Easter communion had been made at St Nectan's. The disagreements within the parish seem to have rumbled on and in 1628 the Bishop of Exeter intervened and ruled that one part of St Winnow parish should attend St Nectan's in the mornings and the other part in the afternoons. This arrangement seems to have gone some way in placating the situation as, according to Canon Miles Brown, the arrangement continued for around 100 years.

At the height of the Civil War around Boconnoc and Lostwithiel in August 1644, St Nectan's was severely damaged: the tower was knocked down and the bells removed, the chapel was slighted and seemingly stood in a ruinous state of limbo until the 1700s. Around this time the tower was capped and the building was in regular use until the late years of the eighteenth century. In 1825 St Nectan's was enlarged by 120 sittings at a cost of £418 11s to hopefully accommodate the extra worshippers which would attend services from among the miners working at the nearby silver vein mine; sadly this upsurge in congregation never materialised. Again, in 1864 the then vicar of St Winnow, Reverend Robert Walker, was instrumental in having a nave built as well as a north aisle, new roof supports, raised gallery and pews installed along with a number of new sittings being added, with monetary help from the Society for Promoting the Building of Church and Chapels.

Over the decades, support for St Nectan's waxed and waned. This was particularly noticeable during the years of the Second World War. It remained in regular use until 1947, being afterwards only used for an annual Remembrance Sunday service. In 1962 the chapel became derelict and the fabric of the building was examined and its future deemed uncertain. After lengthy discussions within the parish it was decided that the Sunday school and the old stables be demolished and the remainder of St Nectan's be saved. Eventually, in 1971 St Nectan's was again opened for worship. Inside are brass memorial tablets to remember the sisters who ran St Faith's Girls' Home, brought to St Nectan's when the home closed in 1950. Also remembered is Sister Susan Mclean, who was for many years a teacher at St Nectan's Sunday school and also a teacher at St Winnow day school. Sometime ago, St Nectan's was the subject of a wildlife survey; surprisingly perhaps there are over sixty trees, shrubs, wild flowers and ferns growing in the churchyard. One of the original church pinnacles is today

Above left: St Nectan's or Nighton's chapel near St Winnow is tucked away amid a maze of narrow roads, and its location will try the map-reading skills of would-be visitors, but the effort is worthwhile!

Above right: Interior of St Nectan's or Nighton's chapel, which exudes a special atmosphere of peace and tranquillity which seems to embrace the visitor!

doing service as part of a style leading from the roadway into the churchyard.

Generally it seems that St Winnow in encased in a tranquil world of its own. There are only the natural country sounds and the gentle voice of the river as it flows by, a peace occasionally disturbed by the sound of a train hurtling past on the opposite bank. St Winnow parish, which seems to be somewhat hemmed in by the great woods of Ethy on the one hand and Lantyan on the other, covers over 6,000 acres including St Nectan's, which at one time was a distinct parish. St Winnoc founded an oratory here on the banks of the River Fowey, a community he organised for a number of years. Eventually he moved to France, where he founded a monastery at Berges-St-Vinoc near Dunkirk and died in AD 717.

St Winnow church dates from the twelfth century and is partly a Norman stone structure. Over the centuries, some parts of the church, notably the south wall have been torn down with an aisle with pillars erected. Restoration took place, and fortunately Mr J. D. Sedding, who undertook the work, was not overzealous; much of the original fabric was retained. Sixteenth-century pulpit bench-ends date from between 1485 and 1630 and there is some fifteenth-century glass. Inside St Winnow church there is a poignant memorial which commemorates Lt Teignmouth Melvill and Lt Coghill of the 1st Battalion of the 24th Regiment, later the South Wales Borderers, who were killed desperately trying to save the queen's colours in the Battle of Isandlwana during the Zulu War of 1879 against the warriors of King Cetshwayo when they overwhelmed the British force of some 1,300 soldiers. They were later awarded the first posthumous Victoria Crosses for their actions. In the churchyard a plain slate tombstone marks the grave of Angela Du Maurier, who was a regular worshipper in the church. Angela Du Maurier was an author in her own right of a number of books. Angela died aged ninety-seven years in London in 2002.

Thomas Lawrence was vicar of St Winnow in 1678 and it fell to him to repair of the church after damage was done to the building during the Civil War. He also restored the screen and generally improved the church. It was during Lawrence's incumbency, July 1714, that the church tower was furnished with its five bells. There is an interesting event, which took place at St Winnow in April 1701, recorded in the church registers. A ship which had been built in a field called Broad Park, part of Newham farm in the parish, was launched. The order for the vessel, weighing 250 tons, was given by Mr Michael Elmes of Tunbridge, who was himself a shipwright. The timbers came from Glynn woods. The vessel was called the *Michael and Frances*.

Among the list of esteemed clergymen in the Diocese of Truro is that of Canon Dr Howard Miles Brown, who was appointed the vicar of St Winnow and St Veep in 1962. At one period Dr Miles Brown worked for the Plymouth Corporation as a civil engineer. Later he entered the ministry and became connected to St Paul's church in Truro. Canon Dr Miles Brown had a lifelong passion for horology and enjoyed making and restoring antique clocks. He published an authoritative book about Cornish clocks and clock-makers in 1961. Canon Dr Howard Miles Brown died in 1994; his tombstone reads 'Priest, scholar and craftsman'.

St Winnow is forever linked with the ever-popular costume drama *Poldark*. It was here at St Winnow on the east bank of the River Fowey, where the swelling river sweeps around a great curve and the Norman church appears to be almost in the river, that Caroline Penvenen married Doctor Dwight Enys. Prior to the filming, various modern-day aspects had to be hidden. Outside electrical installations were disguised by foliage, and peat was strewn on the tarmac church path. Apparently what caused most concern to the cameramen was the likelihood that during the crucial part of the wedding scene, when the bride and groom were leaving the church, a train could be passing on the far bank! Recently a party of elderly *Poldark* fans from Stoke-on-Trent visited St Winnow.

Near the church is a small but fascinating farm museum which will interest all those who are remotely involved with the agricultural community. The museum was started in 1976 by the late Henry Stephens and still continues today. Among the exhibits crammed into a large shed are farm tractors from David Brown, Allis-Chalmers and Fordson. Along with this are a hay bale elevator, a threshing machine, ploughs, a binder, a maize-breaker and an old wagon purchased for the princely sum of £14 in 1909 which was in use until 1965. There are a wide range of smaller tools, saws, scythes, corn measures, butter churns and sundry items to stir the memory. In an adjoining building there is a blacksmith's forge showing bellows, an anvil and tools of the trade.

Leaving St Winnow, the river glides gracefully on toward St Veep, named for another enigmatic saint. No one knows for sure whether the saint was male or female. However, in 1336 Bishop Grandisson of Exeter rededicated the church to St Ciricus and St Julitta. This dedication seems to have been studiously ignored; the present church and parish are always called St Veep. The six bells at St Veep, which were dedicated in 1770, have a fascinating story.

Looking across the Fowey toward picturesque St Winnow church, which has a special peace of its own and will always be linked with *Poldark*, the popular television serial.

The interior of St Winnow church has much of interest, including bench-ends and displays of photographs of the church during a period of restoration.

Canon Howard Miles Brown, vicar of St Winnow, who had a fascination for repairing old clocks. In this picture he is examining a musical, astronomical calendar clock which took him over ten years to build. He is the author of an authoritative book on Cornish clocks and clock-makers.

Originally there were four bells, one of which was cracked and the other three were out of tune. Apparently the vicar at the time, Reverend William Penwarne, together with his church wardens, decided to have the peal recast, adding two bells and 2.25 hundredweight of metal. By all accounts, when the tenor bell was being cast the parishioners collected as much old silver as they could muster and threw it into the furnace to enrich the tone of the bell. Tradition has it that the bells were recast in a field opposite the church. Taken from their moulds, the bells were hung on a girder. By all accounts, Pennington, the bell founder, tapped each bell in turn and leapt for joy! He told those present that it was a virgin or maiden peal, each bell being in perfect pitch, so he was more than satisfied with his work and that the bells can never be excelled. The bells have never been tuned since that time. It is believed that it is the only virgin peal in the entire country and consequently St Veep is a mecca for campanologists.

It is here at St Veep that visitors are least likely to be thinking of Hollywood, Ealing Studios or the professional acting stage, but near to the lych-gate there is a gentle reminder. Surprisingly there is a tombstone, almost unnoticed, standing alongside others, its gold lettering stating that none other than screen idol and heartthrob Eric Portman lies in the quiet earth. For some years Eric Portman loved his Cornish home by a creek of the River Fowey at Lower Penpol cottage, still affectionately known as Portman's Cottage, being the star's private hideaway. Portman was born in Halifax, Yorkshire. In his early years he joined his father, a wool merchant, in the family firm. Later he found employment in the men's department of the Marshall and Snelgrove store in Leeds. At one period he was part of the Halifax Light Opera Company. Later in 1924, he made his professional stage début with the Henry Bayton Co. and then came to the notice of Lilian Baylis, who engaged him to

Lostwithiel and Further Downstream

This old Allis-Chalmers tractor is among many old farm machinery exhibits at the St Winnow Farm Museum.

One of the exhibits which bear the name 'David Brown' at the St Winnow Farm Museum, where the memories of farming in days long gone will be revived.

The blacksmith's shop, once a focal point of most villages, has been carefully re-created in the St Winnow Farm Museum, showing bellows, an anvil and a variety of blacksmithing tools.

St Veep church, tucked away among the trees, has a special claim to fame and is of great interest for campanologists. This view was taken by local photographers Kittow & Son of Fowey more than a century ago.

perform with the Old Vic Company. He also made upwards of forty films between the early 1930s and 1968. Eric Portman seemingly had the happy facility of making all the characters which he played convincing and he gained critical acclaim for many of his performances. Eric Portman's last engagement was in London, where he narrated a documentary on the Brontes for a television film company. In 1948 he won the Ellen Terry Award for best actor. Portman's career also entered the television age, and he appeared in episodes of the *The Prisoner* in the UK and made numerous television appearances in Canada and in America.

Eric Portman was a popular figure in the St Veep area, as a group of mainly elderly ladies who were decorating St Veep church for Easter made quite clear; for them, Eric Portman was definitely 'top of the bill'. According to the *Cornish Guardian* of December 1969, he was a keen supporter of the Lerryn regatta, he patronised the Ship Inn in the village, and spent a short time there on the evening prior to his death. Mr R. J. Phillips, landlord of the Ship Inn at the time, recalled that he mentioned that he was not feeling too well. Mr George Mansell, postmaster at Lerryn, remembered that Eric Portman was 'an amiable sort of chap who kept himself to himself'. However, from time to time, he did make brief public appearances; in June 1969 he opened the Three Arts Festival in the Public Rooms at Bodmin. Eric Portman was ostensibly enjoying his retirement but a close friend of over thirty years said, 'He was putting a brave face on it, but he loved the theatre and was a most dedicated actor.' One thing is certain; at St Veep, he is still well-remembered as a popular and much-respected figure.

Balanced on the opposite side of the river are Lanlivery, St Sampson Golant and Tywardreath. For the most part, Lanlivery, a few miles from Lostwithiel, is a quiet and peaceable place. The parish is surrounded by Luxulyan, Lanivet, Lanhydrock and Lostwithiel and separated from St Winnow, Tywardreath and Golant by the River Fowey. In addition to Lanlivery village, the parish consists of several hamlets, among them Redmoor, Milltown and Sweets House. There is something of a puzzle about the parish name: is it Lanlivery or, as some still say, Lanvorek? The church saint's name dedication too is a mystery. Several names have been suggested but generally, according to the church guidebook, that of St Dunstan is the most favoured.

The church tower, built of granite in the fifteenth century, is 97 feet high, built in three stages, buttressed up to the base of the second stage while the third stage displays eight-sided quoins which rest on corbels showing the heads of angels, lions and humans. The tower is topped off with pinnacles. The main body of the church has arches supported by huge granite pillars. In the east window is an inscription showing that the Prior of Tywardreath was the patron of Lanlivery church. The octagonal font is unusually large and there are monuments to the Kendall family of nearby Pelyn House. Opposite the church door is a large stone coffin, which legend has it once held the remains of a Cornish saint which were brought to Lanlivery from the chapel at Restormel Castle.

From time to time Lanlivery comes up with a surprise. One such time was in February 2003, when an old granite stone which once marked the boundary between Lostwithiel and Lanlivery was unearthed. By all accounts during the Second World War many boundary stones and other roadside stones were removed, as they may give useful information to any invasion force. The stone, probably dating from the first part of the nineteenth century, marked LB (Lostwithiel Borough) on the one side and LP (Lanlivery Parish) on the other, had apparently been buried for around sixty years and has been re-erected at Nomansland.

Above left: Eric Portman, the famous actor who starred in countless films. Portman lived at St Veep for some years. (Photograph from the Rank Organisation)

Above right: This plain tombstone in the churchyard at St Veep remembers Eric Portman, 'actor', and gives no further clue to his illustrious career. Eric Portman made nearby 'Penpoll Cottage' his Cornish home for a number of years.

Tywardreath is justifiably proud of its history. For nearly 500 years from 1088 to 1536 it was dominated by its small Benedictine priory, consisting of one prior and six monks. The first monks came from St Serguis and St Bacchus near Angers in the Loire Valley. The priory complex at Tywardreath probably consisted of chapel, bakery, kitchens, dormitory, refectory and guest accommodation, and would have stood near the site of the present-day church. The monks took vows of silence and never spoke unless someone spoke to them. They spent their time at work on the priory farm, in study performing the priory daily offices, and taking care of the sick and the dying. They never left the priory alone. Tywardreath priory was one of the great landowners hereabouts, covering acres all over Cornwall and stretching into Devon. After the Dissolution the priory, like many others, was virtually 'quarried' by local people for its stone. Seemingly quantities of stone were destined for export back to France, but a vessel foundered and three centuries later the stone was discovered off Pridmouth Bay. Legend has it that some stone was taken to Menabilly. In 1822 the priory site was partially excavated by the then vicar. Tywardreath would thrill

The tower of Lanlivery church is believed to have been as landmark for shipping in St Austell Bay. Inside are memorials to the Kendall family of Pelyn, who were keepers of Restormel Castle from earliest times. This connection ended in 1993 when the last Kendall died. A full-scale church restoration took place in the 1990s.

television's *Time Team* as above ground the priory has almost completely vanished but who knows what and exactly where relics lie below ground!

The earliest church at Tywardreath dates from the fourteenth century and is believed to have been cruciform in shape. Significant additions were made in the fifteenth century when the, battlemented tower and the south aisle were built. On several occasions over the centuries restorations have taken place to either the tower, the roof, the stonework or the interior woodwork. Inside there are several interesting features, including an array of ancient bench-ends and the octagonal granite font. The tower clock was installed to mark Queen Victoria's 1887 Jubilee. Interestingly, at the east end of the churchyard there is a memorial to Bishop Gott, the third Bishop of Truro, who lived at nearby Trenython.

Trenython has a fascinating history. The house was designed and built by an Italian architect commissioned by Giuseppe Garibaldi in 1872. Colonel John Whitehead Peard was an officer in the Cornish Rangers and the band of British volunteers fought to free Italy alongside Garibaldi in his campaigns. Garibaldi never forgot the support which he was given and in consequence, Trenython was given to Colonel Peard as a thank you from the great Italian. When Garibaldi visited Penquite at Tywardreath, the then residence of Colonel Peard, hundreds of people converged on the house to catch a glimpse of the esteemed visitor. Later Trenython was bought by Bishop Gott, the third Bishop of Truro and for the next fifteen years or so it became the bishop's palace. During Bishop Gott's tenure he installed Jacobean wooden panelling in the dining room and also added a private chapel. Eventually another

General view of Tywardreath, which is mentioned in the Domesday Book. The priory, founded in 1092, which grew to be the largest in Cornwall, has almost completely disappeared.

Cherry blossom time at St Andrew the Apostle church at Tywardreath. The church has a fine array of bench-ends and there is a memorial to Bishop Gott, the third Bishop of Truro, in the churchyard.

change of ownership occurred, after the house was purchased by the Great Western Railway. After refurbishments costing around £25,000 it became their seventh convalescent home, accommodating around eighty railway workers. Trenython was officially opened by Viscount Churchill and the house remained as a convalescent home for the next fifty years. Today Trenython is a successful high-class hotel with a worldwide reputation for its cuisine.

The River Fowey glides sedately pass Golant, arguably one of the more attractive places on its banks. In the early centuries it was on the pilgrim track from Wales and Ireland, through north Cornwall and down to the River Fowey and then to Brittany. At one time the quay at Golant was convenient for vessels loading and unloading various cargoes, including coal for the mines, but eventually this came to an end due to the silting up of the river. Boatloads of trippers from Fowey made regular stops at Golant. The railway came to Golant in 1869 though it originally carried china clay down to the docks at Fowey. Later it also carried passengers and Golant Halt was opened in 1896, making for an influx of additional visitors. The viaduct formed Golant Pill by the embankment, making an anchorage for small yachts and boats. Eventually in 1965, the railway reverted back to solely carrying china clay traffic. Golant is a working village, where there is still a core of the original village; the Fisherman's Arms, a quaint street with picturesque whitewashed cottages and a lagoon where small boats seek shelter. In recent years new buildings have sprung up, the village being a favourite place for the retired. A modern feature on the village green is the millennium sundial. At the time of writing a petition is being circulated in an attempt to prevent the construction of a 120-foot-tall wind turbine on high surrounding land, which residents fear will spoil the village.

St Sampson's church stands high above Golant and enjoys expansive views of the River Fowey. It dates from the thirteenth century when it was rebuilt, probably by the monks of Tywardreath priory. The present church was consecrated in the early sixteenth century. Compared with some Cornish church towers, St Sampson's is low and embattled. The pulpit incorporates old bench-ends, as does the reading desk and lectern. There are carved roof bosses, stained glass windows and memorials. According to the church guidebook, Sir John Betjeman, that pre-eminent church-crawler, was not impressed by the pews. He declared them to be the most uncomfortable in the whole of Cornwall. An interesting memorial is that to Reverend J. L. Lyne, who preached his first sermon here in St Sampson's. Between the tower and the porch is the holy well.

Golant has a legend connected with St Sampson and the River Fowey. Seemingly St Sampson lived in a deep, dark cave going back into the hillside by the River Fowey. Tradition has it that when Sampson first entered the cave he came face to face with a fierce, scaly and dragon-like serpent which he had unwittingly disturbed. Angered, the fearsome serpent-like-dragon thrashed about and bit its tail. Then Sampson wound a white girdle from his garment around the creature's neck and from a great height threw it into the River Fowey, where it perished. From that time Golant and the district was freed from the creature's evil influence. Afterwards the cave was revered as the saint's sanctuary. Today the cave is somewhat desecrated with rubbish and is generally out of bounds to casual visitors. It can only be visited with special permission at certain times as it is near the railway line!

Soon the River Fowey embraces the Penpol Creek from the opposite bank.

Trenython House was built in 1908 by Garibaldi as a gift to Col Peard, affectionately known as 'Garibaldi's Englishman'. Once, Trenython was the palace of the Bishop of Truro and later a convalescent home for staff employed by the Great Western Railway.

Patients and nurses find a few spare minutes to pose for the camera at Trenython convalescent home in 1939.

Golant foreshore is a picturesque part of the village, though the rising tide can quickly cover the roadway to catch the unwary.

It is high tide in this view of Golant in 1968.

St Sampson came from Wales in the sixth century and settled at Golant. Later he travelled to Brittany and afterwards became Archbishop of Dol. The present-day church was consecrated in 1509.

7

Downstream to Fowey

At Newton crossroads, a mile or so from Fowey, stands one of the most enigmatic monuments in the area. This is the Tristan Stone, sometimes fondly called the Fowey Longstone by the locals. Dating from the sixth century, it remembers King Mark of Cornwall, Marcus Curnomous. It carries the inscription 'DRUSTANUS HIC IACIT CUMONORUS FILIUS' down the length of the shaft It is believed that it once stood at the great nearby hill-fort of Castledore, which dates from the Iron Age and which structure was later fortified during the Bronze Age. By all accounts it was occupied during Roman times.

There are countless variations of the romance of Tristan and Iseult woven into the legends and history of the area. Certainly the area around Golant is the locale of the Tristan and Iseult tragedy. At one time during the sixth century, Cornwall was seemingly under some kind of debt to the Irish Court, and each year King Mark was obliged to send thirty youths and thirty maidens as sacrifices. His nephew Tristan was so outraged that he challenged the Irish queen's brother, who had the responsibility of bringing the victims to Ireland, and killed him. Tristan too had been wounded by a poisoned sword and only the Irish queen could heal the wound. Later Tristan went to Ireland disguised as a troubadour and he was nursed back to health by the queen and Iseult, her daughter; fortunately Tristan was not recognised as her brother's killer. Once back in Cornwall, it seems that Tristan made a fateful mistake. Apparently he was so struck by Iseult's beauty that he told his uncle about her. Intrigued, Tristan was sent back to Ireland to claim Iseult to be his uncle's bride and queen. By all accounts Tristan and Iseult fell hopelessly in love on the voyage back to Cornwall. Later Iseult did marry King Mark but continued to meet Tristan in secret. Eventually they were discovered by a dwarf, once a court favourite, who betrayed the pair to the king, who condemned the lovers to be burnt at the stake. On the way to the stake Tristan begged to be allowed to enter a small cliff-top chapel to pray. Once inside he barred the door and managed to escape through a small window and drop on to a sandy beach below. Afterwards Tristan rescued Iseult and the couple fled to live in a forest. Later King Mark persuaded Iseult to return to him and forced her to swear on holy relics that she had never been unfaithful to him. On the way to the relics, Tristan, again in disguise, this time as a leper beggar, offered to carry Iseult across a river. The incident served its purpose: Iseult could swear to the king in all honesty that the king himself and the beggar were the only two men who had ever embraced her! Sometime afterwards Tristan went abroad and married another Iseult, daughter of a local chieftain. During a battle in Brittany Tristan was seriously injured and a messenger was sent to Cornwall to beg Iseult to come to him. According to legend the messenger was told to hoist a white flag on the vessel if Iseult

Above left: The famous Tristan Stone, or Fowey Longstone, as it is sometimes called, stands by the roadside at Newtown on the edge of Fowey.

Above right: In August 1960, explanatory plaques were placed by the Tristan Stone by Fowey Old Cornwall Society and were unveiled by Miss Foy Quiller Couch.

was on board and a black flag if she was not. The white flag was duly hoisted but his jealous wife told Tristan that it was a black flag and he died of grief. Iseult too was grief-stricken on learning of Tristan's death and died. King Mark had the couple brought home to Cornwall for burial. Their burial place was originally marked by the Fowey Longstone or Tristan Stone but over the centuries it has been moved. Several eminent historians have raised various claims as to where exactly the stone once stood. In the sixteenth century Leland recorded a broken cross a short distance from Castledore. Borlase, writing in 1769, stated that the cross was found lying in a ditch two bow-shots south from Newton, and it was moved from the Four Turnings on the Fowey to Castledore road. Several decades later Blight records that it was erected near where it lay in the ditch. In the nineteenth century it was moved again and erected at the Four Turnings. Today historians generally agree that it now stands proud by the roadside, a short distance from Castledore. In 1960 a metal plaque gives the transcription of the wording on the original stone, which reads as, 'DRUSTANUS HIC JACIT CUNOMORI FILES – Here lies Tristan son of Cunomorus.' The ceremony at the stone was opened by Mr I. D. Spreadbury and a short history of the stone was given by Mrs I. M. Goddard, after which the green-and-yellow oxidised plaque bearing the inscription, based on the research of Dr Charles Singer, was unveiled by Miss Foy Quiller Couch. The stone was provided by the Delank Quarries at St Breward on Bodmin Moor and the cost was defrayed by Fowey Old Cornwall Society. Recently, December 2012 a storm of protest has erupted in Fowey, after it was revealed that Cornwall Council has given permission for the ancient Tristan stone to be moved, even though it was moved some years ago. This latest move is to allow a, but this time to allow a developer to erect dwellings on a field site adjoining the monument.

At one time numerous wharves and all the associated activities dominated Caffa Mill. Here too there is a slipway for smaller craft and boats are laid up over the winter months. The top section of the obelisk, which remembers Queen Victoria and Prince Albert's visit to Fowey, stands on the quay.

From time to time schemes have been launched to build a crossing of the River Fowey further downstream than Lostwithiel. One such scheme launched in November 1834, if successful, would have made the age-old ferry obsolete. Great consternation must have arisen upstream from Fowey and Bodinnick when it became known that Place, the Treffry estate, was intending to make an application to gain an Act of Parliament which would allow them to erect a suspension bridge over the River Fowey. According to the legal document it was to be located at a spot somewhere between Carne (*sic*) Point and New Quay Cellars, upstream from Caffamill Pill and Bodinnick, though no actual dimensions are given. It was also intended to make proper convenient approaches to the bridge from either side of the river. Tolls would be charged as they were for the ancient ferry crossing. Monies raised would be used for the general maintenance of the bridge and the approach roads. It was also further intended by the power of the Act to (*sic*) 'shut-up, extinguish and annihilate' the rights and usage of the same from the time the approach roads were made open and conveniently passable. The small legal document which bears the official purple ink stamp of the Place Fowey estate was drawn up by Coode's solicitors and dated 12 November 1834. Further research reveals that the bridge was intended to be 500 feet across, the road deck to be 80 feet above high-water level and the end towers to be 120 feet high. One can only assume that the bridge was not built because the Act of Parliament was not forthcoming, or maybe the scheme was withdrawn due to the excessive expenses that it would entail.

Seemingly in March 1858 another crossing scheme was put forward. The *Royal Cornwall Gazette* reports a meeting in the Council Chamber in Fowey to discuss plans for a floating bridge. The meeting was presided over by the Reverend E. J. Treffry of Place and attended by Mr T. J. Austen Treffry, also of Place, Francis Howell of Ethy, Bevil Fortescue of Boconnoc, on behalf of his father, and Reverend J. T. Buller Kitson of Pelynt, who also represented Buller High Sheriff. Subscriptions were raised to £280 for the proposed new road along with a free gift of land. Mr Littleton showed a model of the proposed ferry which he had made, which resembled the steam ferry at Torpoint. The Fowey ferry would be worked by winches and manual labour instead of steam. This bridge, together with the approaches, would amount to over £200, which would be borne by the Hon. G. M. Fortescue, who on the part of Lady Grenville, the owner of the ferry, had liberally proposed a revision and reduction of the present tolls. It appears that about £100 was still to be raised to justify the commencement of the undertaking, and to make an easier approach to the ferry instead of the existing difficult and dangerous one. The scheme was set on foot due to the energetic efforts of the Reverend Kitson of Pelynt. Neither of these schemes came to fruition.

An interesting sight in the river is the pilot boats and the tugboats guiding and moving the giant cargo ships into their position in the river or lying beside the jetties being loaded with thousands of tons of china clay, which has been brought down to the port by rail for further shipment all over the world.

This picture shows *Kelt*, a Dutch-registered vessel of 2,409 tons and 82 metres in length, built in 2008, entering Fowey in May 2012. *Kelt* will be soon loaded with china clay destined for Sweden. Some visitors were surprised at her size, but according to a young lady in the harbour master's office, *Kelt* is quite a small girl compared to some vessels which visit Fowey.

Today the export of china clay is big business, and the loading of the raw material onto these ships is a highly mechanised operation, making for a fast turnaround, as the tide dictates everything, even time and money!

Fowey has always been proud of its lifeboat and its lifeboatmen. Here RNLB Trent Class *Maurice and Joyce Hardy*, the latest in a long line of lifeboats to be stationed at Fowey, rides at anchor in the river, ready to answer a distress call within minutes.

Fowey, the most important port on the south coast, basks in an age-old maritime history, connected with fishing, sailing, trading and shipbuilding. It is recorded that in 1347 Fowey sent forty-seven ships to Edward III to siege Calais. The town is a veritable network of narrow streets lined with quaint buildings and houses which show glimpses of the town's long and proud history. The town was granted its first charter by Prior Theobald of Tywardreath in the thirteenth century. Later Fowey was able to send members to Parliament, which not surprisingly caused some rivalry between the Rashleigh family of Menabilly, who supported the Whigs, and the Treffry family of Place, who favoured the Tories. One thing is certain in Fowey: every street has its gem. The fifteenth-century Ship Inn was once the town house of the Rashleigh family in Lostwithiel Street. The hostelry is popularly believed to have been named after the celebrated *Francis of Fowey*, a Rashleigh vessel which in 1588 sailed against the Spanish Armada. The King of Prussia hotel on Town Quay was in all probability named after Frederick the Great, but popular legend has it that it was named after one John Carter, a notorious Cornish smuggler who was nicknamed the King of Prussia by his friends, and the name stuck. The Sailor's Rest in Fore Street was built in 1889 as a meeting place for seamen while on shore where they could get food, companionship and company until 1950. It is now Old Quay House Hotel. Here too is Noah's Ark, once a sixteenth-century merchant's house. The Fowey Museum in Trafalgar Square, once Market Square, has on view a vast array of historical artefacts, documents and photographs relating to the town and district.

Fore Street, Fowey.

The Ship Inn was once the town house of the Rashleigh family, situated in the centre of Fowey.

General view of the harbour from Town Quay. Ships from all over the world eventually find their way to the port of Fowey. In July 1979 Fowey was the starting point for the twenty-five strong line-up for the Sail Training Association's tall ships race from Cornwall to Port St Mary on the Isle of Man.

The church is dedicated to St Finbarrus. Finn Barr was an Irish missionary from Cork. According to I. D. Spreadbury, around the time of the Norman Conquest the church was already in a state of disrepair and the Normans had it rebuilt. The tower, built between 1457 and 1471, has four stages and soars to 119 feet 1 inch, making it the second-tallest tower in Cornwall. The porch, added in the early sixteenth century, has two open arches and a vaulted roof. The north side of the church once had small cottages crammed against it but they have been removed. Noticeably, the great wall surrounding Place comes within a matter of feet from the church building. Inside the tower has eight bells, a magnificent wagon roof, carved woodwork, fourteenth-century clerestory windows, an early ornately carved Norman font, a pulpit dating from 1601 and a carved screen. The large Perpendicular eastern window features sea stories from the life of Christ, other features including the Rashleigh monument, the Treffry chapel and memorials to other Fowey and local worthies. Over the centuries the church has altered several times. As recently as 1996 the Victorian floor tiles, many of which needed attention, were removed and the floor replaced with Delabole slate.

Place House and the Treffry family, who originally had close connections with Linkinhorne and Callington, have been inexplicably interwoven for the last 700 years into the proud historical fabric of Fowey. Like any one of a number of castles punctuating the course of the great River Rhine, Place House above the River Fowey is a gem. From its unique position on the steep hillside it has commanding views overlooking the bustling town, port and River Fowey, with Polruan and Bodinnick piled on the opposite hillside. Place dates from the

View of The Haven, the home of Sir Arthur Quiller Couch for many years, which property afforded expansive views of Fowey harbour.

Fowey church has the second-highest church tower (after Probus) in Cornwall. Arthur Quiller Couch, later Sir Arthur, married Louisa Hicks here in 1889.

fourteenth century and was built on the site of a former house. Due to the lie of the land Place is difficult to see; only a tantalising glimpse is offered from the massive gate near the parish church, though the visitor is afforded more tantalising glimpses of the house from the river. Richard Carew in his survey of Cornwall published in 1602 described Place as a 'fair and ancient house as a castle builded over looketh the town and haven with a pleasant prospect'. Leland, who stayed at Place in 1538, described it as 'the Glory of Fowey'.

In 1346, during the Battle of Crecy, Treffry captured the French standard and for his courage was knighted on the field. Another John was knighted on the field at Bosworth for his gallantry and support for Henry Tudor. Royal service was evidently in the blood of the Treffry family. In 1500, Thomas Treffry was the builder of Henry VIII's castle at St Mawes and also its first captain. Today Place is the home of the Treffry family, and it is not open to the public. Fowey was vulnerable to attack from the river and in 1457 the French raided the town, rampaged through the narrow streets and attacked Place House. Thomas Treffry was away at court at the time but apparently his wife Elizabeth was made of stern stuff, fighting back against the attackers of her home by pouring boiling lead onto the assailants and eventually repelling them. Soon after the attack Thomas Treffry had the tower built and the walls fortified.

In 1779 the house passed out of the direct line of descent and was settled on two sisters, who sold it to Joseph Thomas Austin, who came into the property through his mother and who remained a bachelor, living at Place with his mother. He assumed the name Treffry in 1836 and created the form of the house which we know today. Starting in 1813 and

Place House at Fowey has been the home for generations of the Treffry family. The house, which overlooks the town and the harbour, is only open to the public by invitation on special occasions.

continuing over the next thirty or so years, he rebuilt Place. Later, in 1820 the distinctive tower just visible from the town, which gave the house its fortified appearance, was built. At this time Place was by all accounts in a poor state of repair. Alterations had been made in earlier years and he was careful to incorporate as much as possible of the old Tudor building into his newest rebuild. The great tower, on the site of that earlier structure defended by the redoubtable Elizabeth Treffry, soars to around 60 feet, affording unsurpassed views of the river and countryside.

A fascinating story is told about the Reverend Edward Treffry who inherited Place in the nineteenth century. The story goes that while Treffry was preaching, news of an impending shipwreck coming in to the shore was being whispered among the congregation in St Fimbarras church. Naturally the congregation became noticeably restless, wishing that they could get to the possible wreck site and grab anything of value that may be washed in on the tide. Apparently, Treffry picked up on this and he too was eager to be there, so the story goes that he cut his sermon short and announced in loud but solemn tones from the pulpit, 'Wait. Don't move.' He then made his way to the church door and said, 'Right. Now we can all start fair.' And leading everyone else, made a headlong dash to the shore.

September 1846 was of great significance in Fowey. The town was a hive of activity and excitement in preparing for a visit of Her Majesty Queen Victoria and Prince Albert, the Duke of Cornwall, and the Princess Royal. The Royal Squadron entered the harbour in mid-morning amid tremendous cheers. The town was dressed over all with flags, bunting and greenery. The barge conveyed the royal party to Broad Slip, which, the *Royal Cornwall Gazette* reports, was tastefully decorated with evergreens and richly carpeted. A royal carriage had been landed from the *Garland* and the royal couple drove off in the company of Mr William Rashleigh of Menabilly and Mr J. T. Treffry of Place. Other gentlemen of the neighbourhood travelled in their respective carriages to Restormel Castle. Obviously Queen Victoria intended to see as much as she could as later the royal party visited the valuable iron mines in the duchy property. The queen and prince, with a lady-in-waiting and three gentlemen, went some 270 fathoms into the mine in a train wagon lined with straw and green baize. Descent into the mine is made by an incline plane, and this is believed to be the first-ever occasion when a queen and sovereign has explored the recesses of a mine, where they inspected the workings on the lode. The prince seemingly took the opportunity to get 'hands on' and broke some ore with a pick. Apparently the queen and prince were so interested that they overstayed their schedule and did not arrive back in Fowey until well into the afternoon. At Place they admired the house both externally and internally. Apparently the prince particularly admired the Porphyry Hall, Jasper and other granites. From Place the royal party returned via the lawn to Broadslip. Peals of bells were rung, music played and several thousand people cheered as the party travelled through the town to re-embark from their ancient duchy to sail to Osborne on the Isle of Wight.

An interesting note relates how the young prince, while on board, mixed with the crew and wore sailor's dress. Seemingly his one and only white duck jacket and trousers had become much soiled by his gambols, and a member of the crew washed and dried the suit overnight; consequently the duke appeared with credit at the morning muster. On hearing what had happened the royal parents were highly amused. One anecdote was reported; apparently a *faux pas* was made while the party were in Falmouth. Seemingly someone inadvertently

A queue of visitors obviously waiting to embark on a pleasure boat trip up the river or out to sea. These excursions have long been popular among visitors to Fowey, and they still are today.

Above left: St Catherine's Castle dates from 1538–42. The tower was built by Thomas Treffry as part of King Henry VIII's southern coast defence system. The castle was restored in the 1850s and fortified during the Second World War; reminders of the defences remain.

Above right: Sir Arthur Quiller Couch, the author of a vast amount of literary work and several popular novels, in his study at home in November 1943. Sir Arthur was mayor of Fowey in 1937 and a great ambassador for the town. (Photograph courtesy Cornwall Centre, Redruth)

Downstream to Fowey

Ships from many countries sail into Fowey Harbour, which is always thronging with river craft.

View of Fowey from the Polruan ferry. Note Place House tucked away among the trees.

referred to the heir apparent as the Prince of Wales. This was immediately picked up by the queen, who said, 'This is Cornwall'; he was always titled not the 'prince' but the 'duke'. Evidently she was not amused! To mark the successful royal visit an 83-foot obelisk with its plinth was erected on Town Quay. Later it became a nuisance to traffic and it was thrown into the harbour. The *Cornish Guardian* of December 1969 records that in 1962 the famous actor Eric Portman, who lived at St Veep, promised £100 to the person who could discover it. It was claimed by a local man but it was decided that the obelisk should remain on the bed of the harbour where it lay off Penleath Point. Today the top section stands at Caffa Mill car park to mark Queen Elizabeth II's Silver Jubilee in 1977.

The importance of Fowey during both World Wars cannot be imagined. At one time large barges were brought to Fowey and anchored upstream of Wiseman's Stone. Vast quantities of ammunition were railed to Fowey and loaded on to barges for shipment. By day men worked on the barges, which were closely guarded by night. Prior to D-Day in 1944, part of the large fleet of vessels destined for Normandy was assembled in the River Fowey. Fortunately, though obviously known to the Luftwaffe, Fowey received only two air raids though there were countless warnings.

Set high above the town is The Haven, home of Fowey's celebrated author Sir Arthur Quiller Couch. Here he lived for several years and took a great interest in Fowey, and was at one time mayor of the town. He wrote a vast amount of work. It was reported in July 1940 that he was working in his garden when a German bomber aircraft dropped a bomb near his house. Another bomb tore away some trees and ripped away part of the cliff, other bombs fell into the sea. On another occasion the ferry was machine gunned as was the coastguard station.

The views across the harbour are expansive and entice the visitor to take the ferry to explore both Bodinnick and Polruan on the opposite side.

General view of Bodinnick by Fowey as it was in 1907. Note the dwelling, once known as Swiss Cottage, later the home of the Du Maurier family called more appropriately, Ferryside. The keen eyed will notice a few subtle changes which have taken place.

8

Bodinnick, Lanteglos-by-Fowey and Polruan

The ferry crossing over the river from Fowey to Bodinnick is only of a few minutes' duration, but if it happens to be to unfamiliar territory, then it can be as exciting as the start of the sea voyage of a lifetime! For over 700 years the ferry has plied the turbulent river crossing. Once called the Horse Ferry, it could carry a coach several horses and the boatmen, with only wide sweeps, found it a struggle to keep the vessel on its rightful course against the strength of the tide. By 1917, with motoring gaining in popularity, a two-car ferry was brought into service. Early motorists must have been pleased as the trip from Fowey to Polruan would have formerly taken them via Lostwithiel; a distance of around 20 miles could now be made in a matter of minutes.

The Du Maurier family holidayed in the Fowey area in the 1920s. Seemingly they instantly fell in love with Ferryside, once called Swiss Cottage, and they purchased the property in 1926 and it still remains in the family. It was on one of Daphne's explorations of the river and surrounding area that she discovered the decaying hulk of the *Jane Slade* at Pont Creek. It was at Ferryside that Daphne began to write her first novel, *The Loving Spirit*, a clever blend of fact and fiction set in nineteenth-century Polruan. The book tells the story of the *Jane Slade* and the Slade family of shipbuilders over several generations. Daphne was given the ship's figurehead and placed it by the front door at Ferryside, where it can be spotted by sharp-eyed passengers crossing the river by ferry.

Today both Bodinnick and Polruan have their individual places of worship; the church of St John the Baptist at Bodinnick was converted from a barn, and there is St Saviour's at Polruan. How the diminutive church of St John the Baptist came about makes for a fascinating story. By all accounts several of the older parishioners had found the journey to Lanteglos church, some distance away from the village, difficult to make and therefore could not attend church on a regular basis. It so happened that the vicar of Lanteglos was delivering parish magazines to the Old Ferry Inn when he asked the landlord, Donald MacFadyen, if he knew of a small room or a plot of land anywhere in the village where he could either hold services or build a small church. There was a nucleus of parishioners who hinted that they would rather like services to be held somewhere in Bodinnick. The landlord immediately offered the use of a room in a cottage and stable belonging to the Old Ferry Inn. Apparently the idea grabbed the imagination of the villagers, who banded together to transform the cottage stable into a church. The post van which had formerly been garaged in part of the building was relocated. The church was named for St John the Baptist because at one time there had been a chapel-of-ease at St John's.

At first everything was somewhat primitive as there was no seating, so the congregation had to bring their own chairs and the only lighting came from storm lanterns. Much of the work

The view of Bodinnick, where the main street is lined with quaint old cottages leading down to the river and the ferry.

Above left: A lone yachtsman weaves his way precariously between the Polruan ferry and other craft as he enjoys life on the Fowey in May 2012.

Above right: Dame Daphne Du Maurier relaxing, probably, at Menabilly, where she lived for a number of years. Menabilly was the template for Manderley in *Rebecca* and featured largely in *The King's General*, arguably her best-loved Cornish novel!

The car ferry arriving at Bodinnick in 2012. The ferry would have been killed off if plans for a suspension bridge over the river had been successful.

Ferryside at Bodinnick, photographed from a River Fowey pleasure craft. Ferryside has been the home of the Du Maurier family since they fell in love with the property in the 1920s. Today Ferryside is the private home of Christian Du Maurier Browning and his family. It was here that Daphne Du Maurier wrote *The Loving Spirit*, her first novel. (Photograph by John Neale, permission of Christian Du Maurier Browning)

Though the general riverscape has changed since this view was taken, some of the buildings, including the limekilns, remain. In the 1960s, author Leo Walmsley lived near here and wrote about the area.

At Bodinnick it is well worth the climb up the hill from the ferry to discover St John's church.

was carried out voluntarily by local people and visitors gave furnishings to the church. The altar cross, hymn board, clergy desk and lectern were all made and given by local men. The stone font, made by a Lostwithiel man, was given by a visitor. The silver crucifix behind the credence table was found lying in the mud on one of the First World War battlefields. The large crucifix above the altar was originally in St Faith's convent at Lostwithiel until that foundation closed in 1950. In 1949 St John the Baptist church was dedicated by Jack Holden, assistant Bishop of Truro, and the first vicar was Canon May. Not surprisingly, today, some sixty and more years on, the church, which must be regarded as one of the smallest Anglican places of worship in Cornwall, exudes a peace and charm of its own which makes it a favourite with countless visitors.

During the Civil War, when Fowey was occupied by the Parliamentary forces a musket was fired from Fowey toward Hall Walk. By all accounts the musket ball narrowly missed King Charles I, but sadly killed a man who was close by.

In September 1948 scores of people from Fowey crammed the Polruan ferry to cross the river in order to witness a ceremony on Hall Walk on the opposite bank. This was the unveiling of two granite memorials by the Bishop of Truro. One was the war memorial to those from Lanteglos-by-Fowey who lost their lives during the Second World War. The other was to honour Sir Arthur Quiller Couch, arguably Fowey's greatest citizen, who loved Hall Walk and Penleath Point. Hall Walk was given to the National Trust by Lt-Col P. F. Shakerley of Tredudwell Manor as a memorial to the war dead of Fowey and Lanteglos and to Quiller Couch. The memorials were designed by Mr J. W. P. Coggin and made at the De Lank quarry at St Breward. The war memorial has a crusader's sword carved down one side and stands near the Bodinnick end of Hall Walk. This memorial was unveiled by Miss M. C. Howell of Crinnis, St Austell, aunt of Lt-Col Shakerley.

The 'Q' memorial is a 13-foot-high incised granite monolith on Penleath Point. The stone has a huge letter 'Q' at the top, the letter being the pseudonym of Arthur Quiller Couch, and a rectangular bronze plaque, with the arms, in colour, of Fowey and Cornwall and those of the Universities of Oxford and Cambridge at the corners, is sunk into its side. Mrs Anne Treffry invited Miss Foy Quiller Couch to unveil the memorial to her father. After the Bishop of Truro had dedicated the memorial, tributes were paid to 'Q' on behalf of Oxford University and also Cambridge University. To signal the end of the ceremony, buglers from the Duke of Cornwall's Light Infantry sounded the Reveille.

Lanteglos-by-Fowey church, apart from a farmhouse and a handful of farm buildings, seems divorced from civilisation but closer scrutiny of the area shows that it is strategic to the population, being midway between Bodinnick and Polruan and central to several hamlets. There seems to be some uncertainty as to the dedication of the named saint. William of Worcester in 1478 favoured St Willow, who was murdered at Lamellin, somewhere near the head of the creek in the valley below. A century or so later Leland discovered a chapel near the head of the creek and mentions it being distinct from the church on the hill above. The church, of which some Norman work survives, is now generally accepted as being dedicated to 'St Wyllow', as Leland wrote in 1548 in the parish of Lanlegiske-juxta-Fowey, being the parish church of Bodinnick and Polruan. Near the end of the fourteenth century a major reconstruction took place which marks the date of the north aisle; the south aisle was added some years later.

By the end of the nineteenth century the church was in a state of near collapse. The tower, the *Royal Cornwall Gazette* reports, was so ruinous that the peal of six bells, locally renowned for their tone, could not be rung. The roof gaped, letting both birds and rain into

The tall granite memorial to 'Q', Sir Arthur Quiller Couch, one of Cornwall's greatest men of letters, stands proudly on Hall Walk. It was unveiled with all due ceremony in 1948.

This view of Lanteglos by Fowey church is from 1904. Bishop John Gott, the third Bishop of Truro, wrote that this church was one of the most interesting in Cornwall!

the building, and the north wall bulged dangerously outwards. The iron-ties which braced the wooden beams and timbers together were rusty and the east window could have done adequate service as a garden frame! The floor, too, was honeycombed with vaults and the whole had to be concreted over, though today marble and slate predominate. Bench-ends from the fifteenth, sixteenth and seventeenth centuries were mixed up with box pews and patchwork of the cheapest carpentry. Fortunately in the nineteenth century Misses Rashleigh and Pinwell carved much of the woodwork and the pulpit we see today. Apparently the walls and arches had been regularly and punctually whitewashed for over forty years.

The church restoration of 1900–6 was a delicate undertaking. The north arcade was leaning toward the aisle and the north wall was pushed outwards by the downward thrust of the roof. Seemingly, however, despite every structural defect St Wyllow's was deemed to be worth saving and to this end the parishioners rallied round the then vicar, the Reverend J. T. Mugford. The arcade was straightened (though it still leans slightly) without the roof being lifted, the tower was also strengthened and the pinnacles, which were lying in the churchyard, have been replaced in their original positions. The cost of the work already undertaken was over £32,000. The restoration programme was overseen by three successive vicars of Lanteglos: the Reverend J. T. Mugford, when he was preferred to Treleigh; Reverend T. O. Wonnacott in his brief incumbency, and later the Reverend C. F. Trustedon, upon whose shoulders most of the burden fell. Some interesting features were discovered, including two recesses in the north wall believed to have once contained figures of saints and two piscinas. Today the interior of the church has some fascinating features.

The ornate square font is believed to be thirteenth-century and of Norman design, the bowl being of Pentewan stone, and it is considered the only Cornish font to be octagonal inside. At one time the Mohun family who came to England with William the Conqueror wielded great influence in the district and were also great benefactors to the church. A little mystery surrounds the tomb of Thomas Mohun, set into the south wall. Thomas Mohun, who died in about 1400, apparently supervised the building of his own tomb. Interestingly, as Thomas was still alive the carver could only put the first two digits of his date of death. It seems remiss that after Thomas' death no one bothered to carve the two missing numerals. There are traces of the original paint which seems to have seeped, fresco-like, into the stonework. Near the altar are the stained glass likenesses of two saints, Priscol and Agnes. Wooden panelling from some of the box pews, bearing the coats of arms of local prominent families, is arranged near the back of the church. The approach to the church was also improved when several steps were built outside the west door and others inside the church to floor level. At the time of the re-opening there was still much restoration work to be carried out.

Outside stands an ancient, four-sided, Gothic lantern cross which was apparently discovered in a trench in the churchyard in 1838. Its ornate carvings shows the crucifixion on one side, the figure of the Blessed Virgin Mary and the Christ child on another, as well as figures of saints. It is believed to be one of the best examples in Cornwall. It was mainly due to the instigation of the Hon. G. M. Fortescue of Boconnoc that it was erected by the south porch. There is also a striking piece of modern metal sculpture in the churchyard by Simon Thomas (RCA) representing the tree of life, donated by the family of W .G.'Gerry' Jenkins. 'Gerry' Jenkins who was born in Wales in 1921 and died in 2001 at Polruan. The sculpture is a form of thanksgiving for his life and a tribute to all those who suffer as prisoners of war, conscience and faith.

Life around the Fowey estuary is governed by the tide. Over the years brides have elected to sail to be married at Lanteglos-by-Fowey. In 1932 it was Daphne Du Maurier who sailed up to Pont Bridge on an early tide and walked the short last leg to the church, where she was to marry Major Browning. Sometimes writers imply that Polruan, lying but a few short minutes away across the estuary from Fowey, is the poor neighbour, but not so; Polruan, clinging to the hillside, a warren of narrow streets with its cottages seemingly placed in picturesque order, has a proud and distinct history of its own.

There is a long tradition of shipbuilding and ship repair, which is still flourishing in Polruan today. Old photographs of Polruan show that fishing nets were draped over large stakes on the quay to dry off in the sun. Apparently after the nets were dry, local ladies took full advantage and hung their washing to dry over the same stakes. Above Polruan is St Saviours chapel, which was seemingly built as a dual-purpose building – as a chapel and as a sea-mark for shipping – and there were bells which could be rung to warn vessels of the dragon's teeth rocks below!

It was in 1922 that Polruan hit the national headlines in an unexpected way. The *Cornish Guardian* reports that the first-ever strike among school pupils in the country took place! The problem was triggered by the appointment of a new headmaster of the boys' school. Although there were three candidates for the post, the management felt fit to appoint Mr L. Tipping from Liverpool. But feelings among the villagers ran unusually high. The school management was blamed for failing to appoint one or other of the two Cornish candidates despite their adequate qualifications. The same newspaper reports in a subsequent week that when the school-reopened after its holidays only half the pupils turned up for lessons! The others went on strike, parading the streets carrying banners and flags of different designs and singing and shouting, 'Down with Tipping' and 'Up Roberts'. By all accounts Mr Roberts was the senior

This view from St Catherine's Castle on the Fowey side of the river glimpses Polruan Point and the open sea.

There have long been shipbuilding and repair yards at Polruan on the east bank of the river. This picture shows one near the foot-passenger ferry terminal. Polruan is 'Plyn' in Daphne Du Maurier's *The Loving Spirit*, her first novel.

assistant at Millbrook, who had served in Salonika, and since his arrival had made himself very popular with the boys and inhabitants. Many people in Polruan thought that he should have been appointed to the post. By all accounts, just before 10 a.m. some of the managers arrived to give Mr Tipping a welcome and then departed. Seemingly from that time until 4 p.m. there was a perfect pandemonium around the premises. The demonstrators were booing and threw stones at the school doors, others were tapping at the windows to distract those pupils inside the class rooms or stood on the window ledges and gave their lungs full play. Under such conditions, the report explains, there could be no order in the school and practically no work was done for the day. Next day a police sergeant from Pelynt was on duty guarding the building from stone-throwing and mud-slinging. The scholars who stayed away from school were again very vocal while they marched through the streets waving flags and throwing stones but they were kept on the move. The public too were sympathetic: they replaced the flags and banners as soon as the police confiscated them from the pupils. However, it appears that Mr Tipping retained his appointment. A further report in the *Cornish Guardian* tells that the only ones hurt in the miserable dispute were the pupils. Seemingly Mr Tipping later came up trumps. He was apparently keenly interested in all things maritime and coached his male pupils in this respect and as a result several took up maritime occupations.

On the afternoon of 19 July 1940, a German bomber aircraft circled the harbour and dropped a stick of six bombs on Polruan. One destroyed a boys' school which had been built in 1878 to accommodate 130 pupils at a cost of around £800, another demolished buildings and others fell into the river or in open countryside. Luckily, school was out and the caretaker had just left the building, so there were no casualties or fatalities. Shortly afterwards the school staff removed everything that could be recycled. Later the boy pupils joined the girls' school, though the classes were separate! On the retirement of the headmaster the school became pupil mixed. The old building was levelled in 1958.

Beyond the quay at Polruan is the ancient blockhouse, which is of great historical significance. It was built of slate and granite, with several gun ports built into the thickness of the walls, which enabled guns to face Fowey while others faced the mouth of the river. According to the Polruan Town Trust information leaflet it is one of the best examples of a chain tower in the country and is balanced on the Fowey side of the river by a similar structure, popularly known as the Fowey blockhouse. It was from the Polruan side that the great chain, with its 16-inch links hung from a huge ring, lay on the riverbed and stretched across the harbour entrance and could be raised in times of emergency. The blockhouse dates from 1380 and its primary purpose was to protect the harbour from attacks by the Dutch, Spanish and French fleets.

On the other side of the coin, the men of Fowey and Polruan were well-known for raids on other shipping and were consequently regular customers of the law courts. Eventually in 1478 the lawlessness of the Fowey Pirates, as they were styled, came to the notice of king. Edward IV (1461–83), who had grown tired of the incompetence shown by local officials in not bringing the local miscreants to full justice. The king imprisoned the officials and the chain was removed to Dartmouth. Ironically, during the Second World War a boom was placed across the harbour between the same two points as the chain in earlier times. Large steel buoys supported a heavy mesh net which hung to the riverbed. The boom could be swung round to allow for the passage of essential shipping. The blockhouse gradually fell into disuse but saw a brief renaissance during the Civil War when the Royalists occupied the building and were able to prevent the Parliamentarian forces from escaping by ship. After the Civil War the importance of the blockhouse faded, but in 1897 Sir Arthur Quiller Couch and a Mr Phelps launched an appeal for funds to rescue the building. Seemingly a surveyor's report recommended immediate remedial work which would cost in the region of £125, most of which was quickly raised. Further repairs were carried out as recently as 1987 by English Heritage, Polruan Town Trust,

The blockhouse on the Polruan side is one of the sights on this stretch of the River Fowey estuary. New access points for the elderly and disabled were opened in 2011 and also steps down to the water's edge.

English China Clays, the Pilgrim Trust and what was then Caradon District Council. Since that time other improvements have been made, including an internal walkway, easier access for the disabled and the elderly and steps down to the water's edge.

Most noticeable on the rocks at the mouth of the river is a white-painted wooden cross on the Polruan side. No one knows exactly when the cross was first erected on the rocks or even why! Leland recorded in 1535 that it was called Pontious Cross, Pontus Cross or Punches Cross. It is generally believed that it originally marked the edge of the harbour and the start of land administered by the priory at Tywardreath. At that time any shipping coming into the harbour had to pay a toll. To this day the cross is still known as Pontius Cross or Punches Cross. Many years ago – it is not known exactly – the local boatmen undertook to paint the cross and to replace any damage caused by storms, an undertaking, incidentally, they still honour! There are also several legends surrounding the cross. Another theory favours the idea that it was erected in the second half of the thirteenth century around the time of Edmund, Earl of Cornwall, and was thought be connected to the now long-ruined St Saviour's chapel above Polruan. One boatman who undertakes to take visitors on short cruises up the river and out beyond the harbour has a wicked sense of humour and puts a modern twist to the story! He, with tongue in cheek, tells his passengers who enquire about the cross that it marks the spot where the last mermaid to visit Fowey was spotted or, alternatively, that it is the site of the last pleasure boat to sink in the harbour – yesterday!

The familiar but enigmatic white cross on the rocks by Polruan is believed to mark the extent of the land once owned by the priory at Tywardreath, and all those using the harbour had to pay a toll.

The River Fowey, whose tiny rivulets rise and are marshalled to form the small stream which originally set the explorer's pulse racing in the remote wasteland of Bodmin Moor, eventually flows between Fowey and Polruan, watched over by their respective sentinels and landmarks guarding the estuary of the river, where it is embraced by the open sea. Having explored the River Fowey from its source on Bodmin Moor, tramped its hinterlands, uncovered some of its closest secrets and mysteries along the way and admired its estuary, is in itself exciting. One thing is assured: the story of the River Fowey is never-ending. It will still be flowing in 1,000 years, but all its secrets and the mysteries of what it has witnessed will never be fully revealed.

Further Reading

Ackland, N. A. and Druce, R. M., *Lanteglos-by-Fowey: Story of a Parish*
Axford, E. C., *Bodmin Moor*
Billington, Phil, *Fabulous Fowey* (Polperro Heritage Press, 2008)
Bray, Peter, *Around and About Fowey*
Brown, H. Miles, *The Book of St Winnow*
Brown, H. Miles, *Battles Royal*
Doe, Helen, *The Jane Slade of Polruan* (Truran, 2002)
Fowey History Group, *Looking at Fowey*
Fraser, Barbara, *The Book of Lostwithiel* (Quotes Ltd, 1993)
Fraser, Barbara, *The New Book of Lostwithiel* (Halsgrove, 2003)
Keast, John, *The Book of Fowey* (Barracuda Books Ltd, 1987)
Mullins, Rosie, *The Inn on the Moor*
Old Cornwall Journal (Federation of Old Cornwall Societies), various issues
Parkes, Catherine, *Fowey Estuary Historic Audit* (Cornwall Archaeological Unit)
Pickering, Isabel, *Images of Yesteryear: Lanteglos-by-Fowey*
Richards, Paul and Reynolds, Derek, *Fowey at War*
Richards, Paul and Reynolds, Derek, *Discover Fowey*

And not forgetting all the chuch guides along the way!

Also by John Neale

Discovering the River Tamar
Following the River Camel